AQA KS3

Science 1

Neil Dixon
Carol Davenport
Nick Dixon
Ian Horsewell

Approval message from AQA

This textbook has been approved by AQA for use with our qualification. This means that we have checked that it broadly covers the specification and we are satisfied with the overall quality. Full details of our approval process can be found on our website.

We approve textbooks because we know how important it is for teachers and students to have the right resources to support their teaching and learning. However, the publisher is ultimately responsible for the editorial control and quality of this book.

Please note that when teaching the *AQA KS3 Science* course, you must refer to AQA's specification as your definitive source of information. While this book has been written to match the specification, it cannot provide complete coverage of every aspect of the course.

A wide range of other useful resources can be found on the relevant subject pages of our website: www.aqa.org.uk.

Although every effort has been made to ensure that website addresses are correct at time of going to press, Hodder Education cannot be held responsible for the content of any website mentioned in this book. It is sometimes possible to find a relocated web page by typing in the address of the home page for a website in the URL window of your browser.

Hachette UK's policy is to use papers that are natural, renewable and recyclable products and made from wood grown in well-managed forests and other controlled sources. The logging and manufacturing processes are expected to conform to the environmental regulations of the country of origin.

Orders: please contact Hachette UK Distribution, Hely Hutchinson Centre, Milton Road, Didcot, Oxfordshire, OX11 7HH. Telephone: +44 (0)1235 827827. Email education@hachette.co.uk Lines are open from 9 a.m. to 5 p.m., Monday to Friday. You can also order through our website: www.hoddereducation.co.uk

© Neil Dixon, Carol Davenport, Nick Dixon and Ian Horsewell 2018

First published in 2018 by
Hodder Education,
An Hachette UK Company
Carmelite House
50 Victoria Embankment
London EC4Y 0DZ
www.hoddereducation.co.uk

Impression number   10 9 8 7 6 5

Year   2022

Cover photo © Laurel/Alamy Stock Photo

Typeset in Vectora 45 Light, 12/14 pts. by Aptara Inc.

Printed and bound by CPI Group (UK) Ltd, Croydon, CR0 4YY

A catalogue record for this title is available from the British Library.

ISBN: 978 1 4718 9992 8

MIX
Paper | Supporting
responsible forestry
FSC™ C104740

Contents

Find the answers at www.hoddereducation.co.uk/AQAKS3Science

How to get the most from this book

Transition

The blue 'Transition' pages cover the required knowledge you need from earlier study. If you are not confident with the content in the 'Core' spreads, or need more help to get started, spend time reading this material and have a go at answering the questions.

Core

The white 'Core' pages deliver the content of the AQA KS3 syllabus for each topic. These pages form the majority of the book and you should spend most of your time studying these pages.

Extend

The green 'Extend' pages go into extra depth about the material you have studied on the 'Core' pages. Have a look at these pages to extend your knowledge and develop deeper understanding of the topic.

Enquiry

The yellow 'Enquiry' spreads are in-depth scenarios and activities to complete in class or at home. They will help you to present data, justify your ideas, devise useful questions and so on.

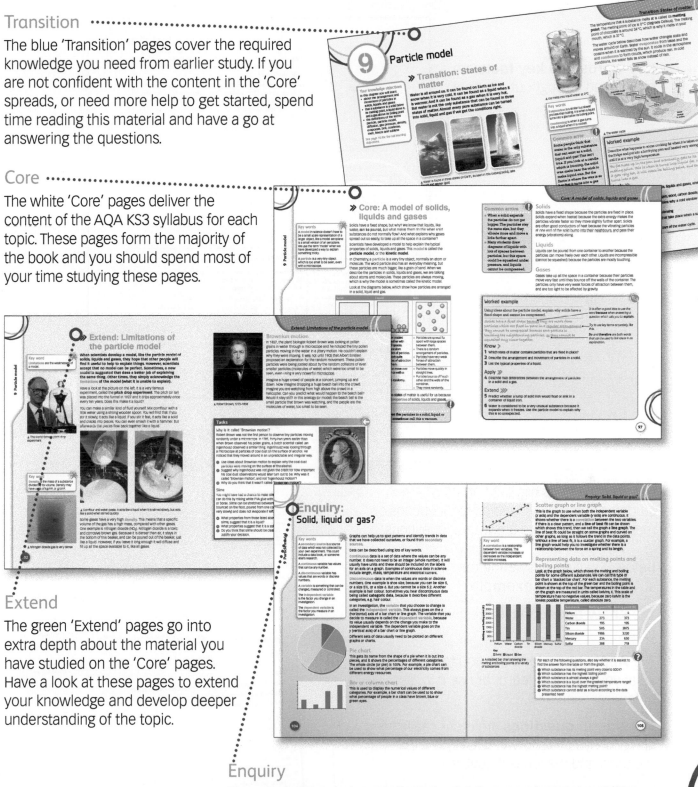

Features you will see

Key words and key facts are highlighted throughout the book.

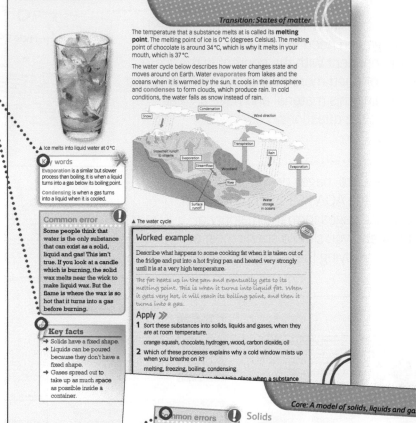

Common errors highlight areas where students commonly have misconceptions or make mistakes when answering questions.

Before you start answering the questions, study the **Worked example**.

The **Know questions** will test your factual recall.

The **Apply questions** ask you to explain what you have just learnt or use it in unfamiliar scenarios.

The **Extend questions** are more challenging and allow you to show you really understand the topic and the ideas it covers.

Enquiry questions are short tasks that can be completed in class or at home. They help you to investigate and process data.

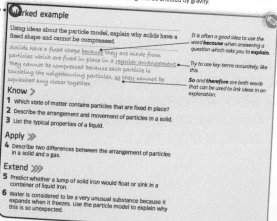

Your Learning objectives

On these pages we have included what are called 'Mastery Goals' from the AQA Syllabus. These are clear statements of what you need to know and how you should be able to apply the content of your course.

At the start of each major section we have created Learning objectives from the Mastery Goals. The Learning objectives are listed under the headings of Knowledge, Application and Extension.

- **Knowledge**: these objectives include important skills that need to be practised for you to become fluent; specific facts that you need to remember, and the important concepts and scientific terms.

- **Application**: these objectives are about more than remembering something or repeating a skill that you have mastered. Application is about using your knowledge and skills and being able to apply them to something new.

- **Extension**: going beyond Knowledge and Application, Extension provides more-challenging Learning objectives. You may need to analyse how two examples are different (or similar); judge if some information is reasonable; or use what you already know to predict what might happen in a new situation.

There are copies of these Learning objectives on our website that you can print off and use as revision checklists.

You will also find the Knowledge Learning objectives at the start of each chapter, to help you see what you will be learning about.

Of course you should always check with your teacher to make sure that you are working with the most up-to-date copy of the AQA KS3 Science Syllabus and the Learning objectives that you are working towards.

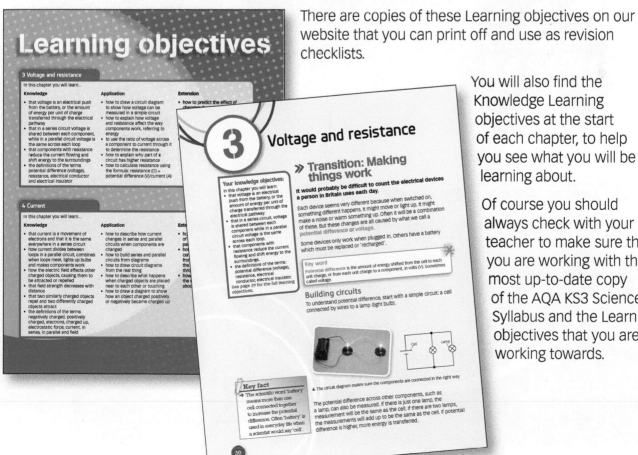

Forces

Learning objectives

1 Speed

In this chapter you will learn...

Knowledge

- about the effect of a resultant force on the motion of an object

Application

- how to calculate the speed of an object using the formula: speed = distance (m)/time (s) or using distance–time graphs
- how to sketch distance–time graphs showing the changing speed of an object and label the changes in motion shown on the graph
- how to describe the effect of relative motion on the speed of an object measured by an observer

Extension

- how to suggest how the motion of two objects moving at different speeds in the same direction would appear to the other
- how to predict how an object's speed will change when the forces on the object change

2 Gravity

In this chapter you will learn...

Knowledge

- about the difference between mass and weight
- that every object exerts a gravitational force of attraction on every other object

Application

- how to draw a force diagram for a problem involving the force due to gravity
- how to describe how the gravitational force of attraction changes with distance and mass
- how to calculate the weight of an object using the formula: weight (N) = mass (kg) × gravitational field strength
- how to calculate and compare a person's weight on Earth with their weight on different planets

Extension

- how to draw conclusions using data about orbits, based on how gravity varies with mass and distance
- how to suggest implications for space missions of the variation of gravity

Speed

» Transition: On the move

Look around and you will see lots of different things moving. Cars and buses on the road. People walking down the street. Runners in a race. The International Space Station orbiting overhead.

▲ Measuring how quickly runners complete a race is very important in sport

To measure **speed** we need to know how far the object has moved, and how long it took to travel that distance.

A speed camera takes two photos of a moving car, with 0.5 s between the photos. The marks on the road allow measurement of the distance the car has travelled.

Comparing speeds

The speed of an object tells you how fast it is going. The faster an object is moving, the quicker it will travel a certain distance. Most people can walk about 5 km in 1 hour. Snails have been measured travelling a distance of about 1 m in 1 hour. People can walk 5000 times faster than snails. Sometimes we need to use different timescales to measure movement. Geologists measured the speed of the Fox Glacier in Antarctica. The glacier only moved a distance of 2 km in 1 year.

Key word

Speed How much distance is covered in a certain time.

▲ The marks on the road are used to measure the distance the car travels

▲ Tatyana was the winner of the race because she travelled the race distance in a shorter time than Manuela

⭐ **You can read more about forces in Pupil's Book 2, Chapter 1.**

Common error !

In everyday life, moving objects often slow down and stop. Some people think there are no forces acting on them. However, there is a force due to friction. That is why the motion we see changes.

Forces and motion

If there are no forces acting on a moving object it will keep moving until a force is applied. Objects will only change their motion when there is an unbalanced force acting on them.

In ice hockey, once an ice hockey puck is moving, it will keep going at a steady speed until a force is applied. The force might be due to the puck bouncing off the side wall of the ice rink, or being hit by a hockey stick.

Key facts

→ The higher the speed of an object, the shorter the time taken for a journey.

→ Forces change the motion of an object. An applied force can speed up or slow down the object, or change the direction the object is moving in.

Common error !

When comparing the speed of two objects, sometimes people only look at the number. To correctly work out which object is moving faster, the unit is also important. To compare objects you need to use the same units for speed for them both.

Worked example

Ali can walk 5000 m in 30 minutes. If her speed stays the same, how far can she walk in 1 hour?

The time Ali is walking for has doubled, so the distance must also double. She will walk 10 000 m.

Apply »

1 Put the objects in the picture in order of how fast they move, with the slowest first.

2 Two lorry drivers are making a delivery. They both have to drive 48 km to a town. Sam travels at an average speed of 40 km per hour and Marie travels at an average speed of 50 km per hour. Who will get there first? Explain why.

3 An eagle flies 120 m in 10 s. How far will it fly in:

a) 20 s?

b) 100 s?

c) 5 s?

4 Ali can jog twice as fast as she walks. If she can walk 5000 m in 30 minutes, how long will it take her to jog 5000 m?

» Core: Calculating speed

To work out the **average speed** something is moving, we need to know how far it has travelled, and how long it took to travel that distance.

We can use the equation:

$$\text{average speed} = \frac{\text{distance travelled}}{\text{time taken}}$$

▲ Amal walking to the post box

Worked example

A pigeon can fly 60 m in 3 s. What is the average speed of the flying pigeon?

$$\text{average speed} = \frac{60\,m}{3\,s} = 20\,m/s$$

The units for speed depend on the units used for distance and time. If the distance is in metres (m) and the time in seconds (s) then the speed is measured in metres per second (m/s).

For slower speeds, we can use units of metres per hour (m/hour) or even metres per year (m/year). If you're travelling in a car, you might use miles per hour (mph).

Finding speed from a graph

For some journeys the speed might be different at different times. Average speed doesn't show this.

We can use a graph to tell the 'story' of a journey and show the different speeds at different times. Distance travelled is plotted on the vertical axis, and the time it took to travel this distance is plotted on the horizontal axis.

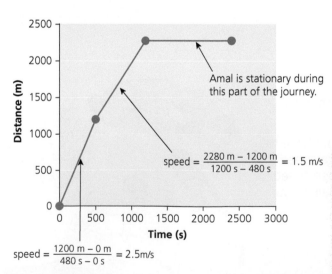

Amal is stationary during this part of the journey.

$$\text{speed} = \frac{2280\,m - 1200\,m}{1200\,s - 480\,s} = 1.5\,m/s$$

$$\text{speed} = \frac{1200\,m - 0\,m}{480\,s - 0\,s} = 2.5\,m/s$$

▲ The graph shows the story of Amal's journey to the park. Her speed changes during different parts of her walk

Amal leaves home and walks for 480 s to reach the nearest post box, which is 1200 m away. She then meets up with her friend and continues to walk for 720 s to the local park, which is 1080 m away. They find a bench and sit and talk for 1200 s. The graph represents Amal's journey.

The gradient of a distance–time graph can also be used to calculate an object's speed.

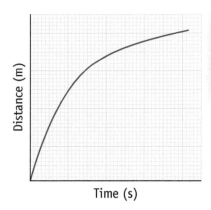

A car is driving above the speed limit. The driver notices a speed camera by the side of the road ahead and puts the brakes on. She slows down to below the speed limit. The distance–time graph for her journey is shown in the graph below.

In this graph the line is curved, and the gradient decreases. The speed of the object is changing and getting slower. The object is moving less distance in the same amount of time.

Key facts

→ A horizontal line on a distance–time graphs shows that the object is not moving at that point in time.

→ A straight diagonal line on a distance–time graph shows that the object is moving at a steady speed.

→ The gradient (or slope) of a graph shows how steep the line is. It is calculated by dividing the change in the y-axis value by the matching change in the x-axis value.

Worked example

Teresa

Salim

▲ Teresa and Salim are pushing the car. It starts to move slowly, but speeds up as they push

The graph shows the distance–time graph for the car that is being pushed by Teresa and Salim.

(a) Describe what is happening to the speed of the car between 0 and 4 s.

(b) Calculate the speed of the car between 5 and 10 s.

(c) Calculate the average speed of the car.

(d) Explain why your answers to (b) and (c) are different.

(a) The speed of the car is increasing, because it moves a bigger distance every second.

(b) $speed = \dfrac{distance}{time} = \dfrac{12.5\,m}{5\,s} = 2.5\ m/s$

(c) $average\ speed = \dfrac{total\ distance}{total\ time} = \dfrac{17.5\,m}{10\,s} = 17.5\ m/s$

(d) Between 0 and 5 s the car starts from 0 m/s and slowly increases in speed. This makes the average speed lower than the steady speed between 5 and 10 s.

Know >

1 What two quantities do you need to know to calculate the average speed of a cyclist?

2 Beth has found out that bamboo plants can grow at 10 cm per day. The willow tree in her garden grows 1.8 m in 4 months. How can Beth compare the growth rate of the two types of plant?

3 How does a distance–time graph show:

a) a steady speed?

b) an object that is not moving?

Apply >>

4 A robin flies 80 m in 10 s. What is the robin's average speed?

5 The two hot air balloons shown both set off at the same time. One balloon travels 32 km in 2 hours before landing. The other balloon travels 11 km in 1 hour. Calculate the average speed for each balloon.

6 Usain Bolt ran 100 m in 9.58 s. Calculate his average speed.

7 Holly cycles to Luke's house. She waits until Luke is ready, and then they cycle to school together. This is the distance–time graph for their journey.

▲ How fast hot air balloons travel depends on the wind speed and wind direction

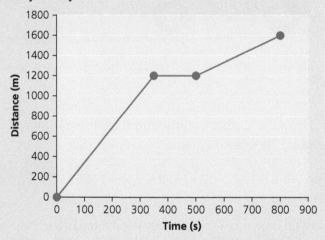

a) Calculate Holly's speed as she cycles to Luke's house.

b) How long does Holly wait for Luke?

c) Calculate Holly and Luke's speed as they cycle to school.

d) Calculate Holly's average speed for the whole journey.

Extend >>>

8 André was in a bike race. His average speed was 11.5 m/s. He finished the race in 9788 s. Calculate the distance André cycled during the race.

9 UK speed limits are usually given in miles per hour (mph). 1 mile is equal to 1609 m. An ambulance is driving on a motorway at 85 mph. What is this speed expressed in m/s?

10 Every autumn, Gray whales travel about 11 000 km so that they can have their young in warmer waters. Gray whales can swim at an average speed of 2 m/s. Calculate how long it takes the whales to travel to the warmer waters.

11 Sketch distance–time graphs to show the following:

a) a runner travelling at a steady speed

b) two race horses that start from the same place, but one is moving faster

c) a child walking from their house to a local shop, waiting at the shop for a while, and then jogging home.

d) a car slowing down over a speed bump, and then speeding up again.

Enquiry »»»

12 A sports scientist measured the effect of wind on how fast athletes could run a 100 m race. She measured the time it took to run 100 m when the athlete was running into the wind. This table shows five different wind speeds (in m/s).

Wind speed (m/s)	1	2	3	4	5
Race time (s)	11.06	11.12	11.19	11.27	11.35

a) Plot a graph of race time (on *y* axis) against wind speed (on *x* axis).

b) Describe the pattern shown by your graph.

c) Use your graph to estimate how long the athlete would take to run 100 m if the wind speed was 0 m/s (i.e. no wind).

» Core: Relative speed

If you are standing by the side of a road, a car travelling at 70 mph goes past very quickly. However, if you're sitting on a train travelling at 100 mph going past the same car, then you will overtake it. It will appear to move backwards.

▲ An uncrewed Dragon spacecraft attached to the International Space Station for docking

The International Space Station (ISS) orbits the Earth at over 7600 m/s. There are usually three or six people living on the ISS at any one time. Every few months, new supplies are flown to the ISS by uncrewed spacecraft such as Dragon.

To safely dock and connect, the ISS and the spacecraft have to travel at the same speed. This means that their **relative motion** is zero.

The relative speed of objects depends on their relative motion. Calculating relative speed depends on whether objects are moving in the same direction or in opposite directions.

Relative motion	Calculation
Objects moving in the **same** direction	Fastest speed – slowest speed
Objects moving in the **opposite** direction	Fastest speed + slowest speed.

Key word

Relative motion Different observers judge speeds differently if they are in motion too. An object's speed is relative to the observer's speed.

Acceleration

If we measured Libby's speed on the slide, every second she would go a little bit faster than before. Her **acceleration** shows how quickly her speed increases.

Key word

Acceleration How quickly an object's speed increases or decreases.

When a driver presses the brakes in a car, the car's speed will reduce every second until it stops. This is often called deceleration.

Worked example

Two lorries are travelling on a motorway. The yellow lorry is travelling at 55 mph. The red lorry is travelling at 58 mph. Calculate the relative speed of the red and yellow lorries.

The two lorries are moving in the same direction therefore:
Relative speed = fastest speed – slowest speed
= 58 mph – 55 mph = 3 mph.

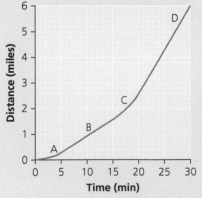

Apply >

1 Libby is on a slide that is 8 m long, and it takes her 5 seconds to reach the bottom. Calculate Libby's average speed as she goes down the slide.

2 This is the distance–time graph for Josh's cycle to school.
Which section of the graph, A, B, C or D, shows where Josh was:

a) travelling at a high steady speed?

b) travelling at a low steady speed?

c) accelerating?

3 Asha and Desiree are running a 4 × 100m relay race. Asha has the baton and is running at 9.6 m/s. How fast should Desiree run so that the relative speed between her and Asha is 0 m/s?

4 Rhiannon and Areti are running a marathon. They both pass a checkpoint at the same time. As they go past, Rhiannon is running at 2.3 m/s and Areti is running at 2.5 m/s. Calculate the relative speed of the two runners to each other.

5 Pavel is throwing a ball for his dog. He throws the ball at a speed of 15 m/s. The dog runs after the ball at a speed of 20 m/s. Calculate how fast the ball is moving relative to the dog?

6 Oliver is standing at the top of a down-escalator. Amy is standing at the bottom of an up-escalator. Both escalators are moving at 0.75 m/s. Calculate the relative speed of Oliver and Amy as they pass each other.

Extend >>

7 Joe is walking at a speed of 1 m/s along a moving walkway at the airport. The surface of the walkway is moving at a speed of 0.5 m/s in the same direction Joe is walking. Kylie is standing still by the moving walkway. Calculate how fast Joe appears to be moving from Kylie's point of view.

8 A 5 m long tractor is driving on a dual carriage way at 40 km/hr. A motorcaravan starts to overtake the tractor at a speed of 42 km/hr. The motorcaravan needs to travel a distance of 24 m past the tractor before it can safely pull in front. Calculate how many seconds it takes the motorcaravan to overtake the tractor and pull in.

Enquiry >>>

9 Alex is rolling a ball down a ramp. Alex thinks that if the height of the ramp is increased, the ball will be moving faster as it reaches the bottom of the ramp. Describe the measurements Alex would need to take to test his hypothesis. Identify the independent and dependant variables in the experiment.

> **Key word**
>
> **Hypothesis** An explanation you can test which includes a reason and a 'science idea'.

» Extend: Bloodhound SSC

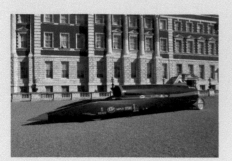

Bloodhound SSC is a supersonic car which is designed and built to set a world land speed record of 1000 mph (447 m/s). The land speed record of 763 mph was set in 1997.

The Bloodhound project started in 2008, and many different people have been involved with the design of the car. Some of the different types of careers in the project include:

- materials scientist
- aeronautical engineer
- design engineer
- electrical engineer
- computer scientist and programmer
- journalist
- racing driver
- graphic designer.

★ **You can find out more about Bloodhound SSC on their website: www.bloodhoundssc.com**

Tasks

1. Use a careers website to research one of the careers involved in the Bloodhound project. Explain why someone with the career you have researched would be needed in the project.

Computer scientists have used computer models to calculate the motion of the car. The table shows a simplified set of data for one run of the car.

Using computer models lets the engineers test different variables to see how they might affect the motion of the car.

Time (s)	Distance (m)
0	0
17	800
25	1700
29	2800
35	5000
42	7600
47	9800
52	11 700
59	13 500
69	15 000
97	16 100

2. Suggest why the Bloodhound engineers are using computer models to help them design the car.
3. Plot a graph of distance (on the *y* axis) against time (on the *x* axis).
4. Calculate the average speed of Bloodhound over the whole journey time.
5. To break the land speed record, Bloodhound will have to travel at 447 m/s for at least 1 km of the journey. Use your graph to evaluate if the car would break the land speed record using this model of its motion.

The Bloodhound team, with the help of a geographer, also had to find the right place for the record run to take place. They used Google Earth to search the world for a large area that could be used as the record run track.

There are a number of factors that the team looked for:

- perfectly flat and smooth area
- at least 19 km long and 5 km wide
- no plants or trees nearby
- dry weather for a large part of the year
- accessible for large trucks.

▲ Even communicating with the driver will be difficult at 1000 mph. The Bloodhound team are testing their radio systems at speeds of 500 mph to start with.

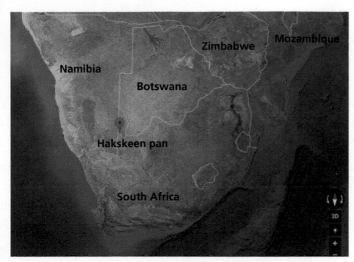

▲ The run location chosen by the team is a mud flat in the Kalahari desert in South Africa. It is called Hakskeen pan.

6 Suggest why the team have chosen a desert as the run location.

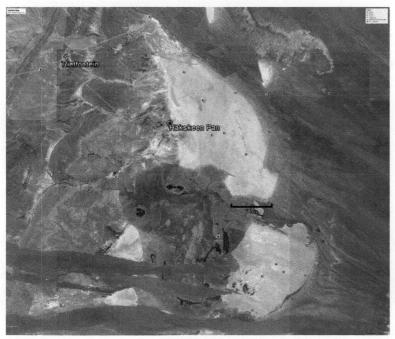

7 This is a map of Hakskeen pan.
  a) Use the scale bar to measure the size of Hakskeen pan.
  b) Use your answer to (a) to decide if Hakskeen pan is big enough to be the record run track.

Enquiry:
What affects the speed of a toy car on a slope?

Child's play

If you put a toy car at the top of a slope, it will start to move down the slope. This effect can be used to make the toy shown on the left. The toddler puts the car at the top of the ramps, and it moves along. At the end of each ramp, the car 'jumps' onto the next one. The car speeds up as it goes down each ramp. This is quite exciting when you're two.

The angle of the slope will change the speed of the car on the ramp. This will affect how much fun the toy is to play with. If the slope is too shallow, the car moves slowly and isn't much fun. If the slope is too steep, the car moves too fast at the end and can 'overshoot' the next ramp.

❶ Name **two** other factors that might affect the speed of a toy car at the end of a slope. **❓**

Measuring speed

We can use a simplified model of the ramp racer toy to investigate how the angle of the slope affects the speed of the car.

Card   Light gate

Method 1: Using average speed.

A stop watch is started as the toy car is released at the top of the ramp. When the car passes a point a known distance down the ramp the timer is stopped. The average speed is equal to distance/time.

Method 2: Using speed at the bottom of the ramp.

A 10 cm long piece of card is attached to the car. A light gate is put at the bottom of the ramp and connected to a datalogger. The datalogger starts timing when the card blocks an infrared beam of light across the light gate, and stops timing when the beam is detected again.

② Give one advantage and one disadvantage for each method of measuring the speed of the toy car.
③ Explain how the datalogger would use the data to calculate the speed of the car.

Gathering data

Carol decided to use method 2. She used the equipment shown to measure the speed of a toy car at the bottom of a ramp. The data are given in the table.

Angle	Run 1 speed (cm/s)	Run 2 speed (cm/s)	Run 3 speed (cm/s)	Mean speed (cm/s)
2°	24.106	24.120	24.211	
3°	46.367	44.840	45.032	
4°	61.823	61.773	61.485	61.694
5°	73.597	74.375	63.487	73.919
6°	83.451	83.921	83.498	83.623
7°	92.748	92.340	93.033	92.707
8°	100.731	100.622	100.811	100.721
9°	107.635	107.496	107.002	107.378
10°	114.462	115.315	114.393	114.723
11°	120.728	121.772	120.866	121.122
12°	126.834	126.218	126.788	
13°	133.216	132.696	133.204	

④ Suggest what factors Carol kept constant when taking the data.
⑤ Why did Carol take three measurements of speed for each angle?
⑥ For the measurements at 5°, the third run speed was 63.487 cm/s. Carol did not use this result and took another measurement at the same angle. Explain why she did this.
⑦ Complete the table by calculating the values for the missing mean run speeds.
⑧ Plot a graph of speed (on the *y* axis) against ramp angle (on the *x* axis).
⑨ The datalogger and light gate can measure the time to 1000th of a second (three decimal places). Discuss whether this is a useful number of decimal places for plotting your graph.
⑩ Use your graph to describe how the speed of the toy car changes as the angle of the ramp increases.

2 Gravity

» Transition: Falling down

Your knowledge objectives:

In this chapter you will learn:
- about the difference between mass and weight
- that every object exerts a gravitational force of attraction on every other object

See page 7 for the full learning objectives.

Hold a paper cup in mid-air. Let go. What happens? The cup will fall to the ground. From being very young we learn by trial and error that unsupported cups (and other objects) will fall towards the ground.

There is a force due to gravity between the cup and the Earth. It is an attractive force, so the cup is attracted to the Earth, and falls towards it.

An object falling helps us to decide what is 'up' or 'down'. No matter where you stand on Earth, objects will always fall 'down'.

It's not just the Earth though. The force of gravity attracts every object to every other object. If you're reading this book sitting next to someone, then there is an attractive force due to gravity between you both. However, gravity is quite a weak force, so you don't notice the attraction normally.

Gravity is a **non-contact force**. An object does not have to be in contact with the Earth for its motion to be changed by the force of gravity.

▲ Young children learn very quickly that if they let go of a toy it will fall down

Key word

Non-contact force is a force that acts without direct contact.

Common error

Some people think that only very big objects have a gravitational force. However, there is a force due to gravity between objects even if they are small. Every object attracts every other object.

The strength of gravity

The force of gravity increases as the mass of an object increases. The force of gravity decreases as the distance between objects increases. The further away from the Sun a planet is, the smaller the force of gravity acting on it.

Isaac Newton was a mathematician and scientist who lived in the 17th century. In 1687 he published a book called *Mathematical Principles of Natural Philosophy*. In the book he wrote about a theory of gravity.

There is a story often told that Newton created the theory of gravity while sitting under an apple tree and an apple fell on his head. The story is probably not true.

▲ People on the other side of the world from you don't fall off, because gravity also pulls them towards the centre of the Earth

▲ Isaac Newton

However, Newton did realise that the force that made apples fall from a tree was the same force that kept the Earth in orbit around the Sun.

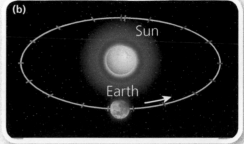

▲ Gravity acts on apples (a) and planets (b)

Know >

1 What type of force is gravity?

Apply >>

2 Here is a simplified diagram of rain clouds above four places on Earth. Copy the diagram into your book.

a) Draw an arrow from each cloud to show the direction of gravity acting on the rain.

b) Explain why you have drawn the arrows in this way.

3 Patrik puts a cup on the table. He says 'Gravity is not acting on the cup because it is not falling.' Explain why Patrik's statement is wrong.

4 The Voyager 1 spacecraft is over 20 000 million km from the Sun. It has been travelling in space since 1977. Describe how the force of gravity on Voyager 1 has changed during its journey.

Cloud

Cloud   Earth   Cloud

Cloud

▲ Voyager 1 is now travelling in interstellar space after exploring the solar system. You can find out more about the Voyager spacecrafts at http://voyager.jpl.nasa.gov

▲ Anne has climbed Mount Kilimanjaro. The force due to gravity on Anne will be slightly less at the top of the mountain than at the bottom

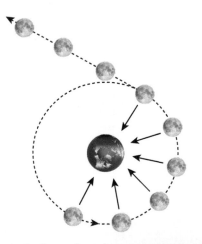

▲ The force of gravity constantly changes the direction of motion of the moon. It would keep going in a straight line if it wasn't for pull of the Earth's gravity.

Key facts

→ Gravitational field strength is given the symbol, g.
→ On Earth, $g = 10\,\text{N/kg}$. On the Moon, $g = 1.6\,\text{N/kg}$

» Core: Mass and weight

In everyday life we use the words **mass** and **weight** to mean the same thing. In scientific terms they are not the same.

Mass is a property of the object. As long as the object doesn't change, the mass will stay the same wherever the object is.

The weight of an object is a measure of the force due to gravity acting on the object. The weight of an object can change, even if the mass does not change. The weight depends on where an object is, and what other objects are around it.

The weight of an object can be calculated using the equation:

weight (N) = mass (kg) × gravitational field strength (N/kg)

$$W = m \times g$$

Gravitational field strength depends on the mass of the planet and how close you are to it.

• The larger the mass of the planet, the bigger the value of g.

• The closer you are to a planet, the bigger the value of g.

The value of gravitational field strength decreases as you get further away from the surface of the Earth. Mountain climbers would weigh less at the top of a mountain than they did at the bottom of a mountain.

There is a gravitational **field** around all massive objects. This is the area in which other objects will feel the force of gravity.

Know >

1 Name the force that keeps the International Space Station orbiting the Earth.

2 What two things do you need to know to calculate the weight of an apple?

3 How would the gravitational field strength on the surface of the Earth change if the Earth's mass increased?

Common error

Many people think that there is no gravity on the International Space Station (ISS). In fact, the value of g on the ISS is only about 11% less than the value on the Earth's surface. However, because both the ISS and the astronauts are falling around Earth at the same rate, the effective force due to gravity acting on them makes them appear weightless.

Worked example

Yutu is a Chinese lunar rover. It landed on the Moon on 14 December 2013. *Yutu* is Chinese for 'Jade Rabbit'.

On Earth, the mass of the rover was 140 kg.

a) What is the weight of the rover on the Earth?

b) What is the mass of the rover on the Moon?

c) What is the weight of the rover on the Moon?

Yutu on the surface of the Moon

a) Use weight = mass × gravitational field strength
   On Earth g = 10 N/kg
   so weight = 140 kg × 10 N/kg = 1400 N

b) The mass of the rover doesn't change on the Moon.
   It is still 140 kg.

c) On the Moon g = 1.6 N/kg
   so weight = 140 kg × 1.6 N/kg = 224 N

Apply >>

4 Calculate the weight of a 100 g apple.

5 A sofa weighs 400 N. What is the mass of the sofa?

6 Georges has a mass of 54 kg. What is his weight?

7 Suggest how you could lose weight without losing mass.

8 Copy and complete the table to show your weight* on another planet.

Planet	Gravitational field strength, *g*, on the surface (N/kg)	Weight (N)
Mercury	3.7	
Venus	8.9	
Jupiter	23.1	
Uranus	8.7	

*If you don't know your mass in kg then use a value of 45 kg.

Extend >>>

9 The force of gravity acts between any two objects. If the mass of one of the objects is doubled, then the force of attraction due to gravity between the two objects also doubles. Predict what will happen to the force due to gravity between two objects when:

a) the mass of one object is increased by 4 times

b) the mass of one object is halved.

10 Use the information for the gravitational field strength on different planets given in the previous table to put the named planets in order of mass.

Enquiry >>>>

11 When you put ingredients into the bowl of a set of kitchen scales, the downwards force due to gravity moves the scales to show the weight of the object. The scale is marked in kg.

a) What is the correct unit that should be marked on the scale? Explain why.

b) Anne took a set of kitchen scales to the top of a mountain. She used the scales to measure out the ingredients for a meal. Explain why there may be more of each ingredient than Anne expects.

▲ Scales measure the weight of ingredients, not their mass

▲ Fridge magnets don't fall down when placed on a magnetic surface

» Extend: Comparing gravity and other forces

Gravity is a non-contact force. Magnetism is another example of a non-contact force.

When we put a magnet onto a ferrous metal surface, such as a fridge, it stays where it is placed. Although there is a force due to gravity acting on the magnet, the force is not strong enough to pull the magnet to the ground.

A simple fridge magnet lets us compare the relative strength of gravitational and magnetic forces. It shows that magnetic forces are larger than gravity.

The magnetic force can be attractive or repulsive. Gravitational force is only attractive.

Contact and non-contact forces

Non-contact forces are all due to fields. Objects in the area feel a force due to the field.

Contact forces only occur if objects are touching. Contact forces such as friction and air resistance oppose the movement of an object.

The table compares different forces that you may have already met.

Name of force	Type of force
Gravity	Non-contact
Electrostatic (static electricity)	Non-contact
Magnetic	Non-contact
Friction	Contact
Drag/air resistance	Contact

Anti-gravity and levitating frogs

Gravity is an attractive force. It always acts to pull objects together. This is different from magnetism, which can be attractive or repulsive.

The idea of anti-gravity machines is often found in science fiction. So far, science has not found a way to create a repulsive gravitational force. However, physicists have used other forces to balance the effect of gravity. This makes objects 'levitate'.

Professor André Geim and colleagues used a very strong magnetic field to make a live frog levitate. The image shows the frog floating in mid-air in the middle of a cylindrical magnet.

Key fact

→ André Geim is the only person to have won both an Ig-Nobel prize (for his work on levitating frogs) and a Nobel prize (for his work on graphene).

▶ A frog floating in the middle of a very strong magnet. You can find out more about this experiment at www.ru.nl/hfml/research/levitation/diamagnetic

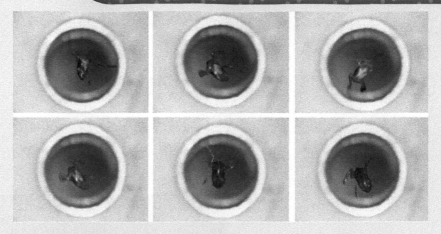

This is not really anti-gravity, because gravity is still acting on the frog. However, the force due to gravity is balanced by the force due to the magnetic field so the frog floats.

Task

❶ Draw a force diagram to show the size and direction of the two forces acting on the frog as it floats in the magnet.

Satellites

Artificial satellites are a key part of life today. There are over 4200 satellites in space. Only about 1400 of these are still working. The rest are broken, or have been shut down.

The diagram shows two types of satellite orbit. Both types of orbits are circular.

The table below gives details of three different satellites currently orbiting the Earth. Other satellites that orbit over the equator are used by military organisations. That means that details about the orbit and mass of the satellites aren't available.

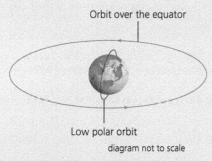

Orbit over the equator

Low polar orbit

diagram not to scale

▲ Satellites orbit the Earth at different distances

Satellite name	Type of orbit	Used for	Distance above surface of the Earth (km)	Mass of satellite (kg)
Iridium	Low polar orbit	Communications	781	689
MetOp A	Low polar orbit	Weather satellite	817	4093
Inmarsat-3	Orbit over the equator	Communications	35,786	2068

Task

❷ Use the information in the table to decide if the following conclusions are supported by the data. Explain your answer for each conclusion.
Conclusion 1: Communication satellites need to be close to the surface of the Earth.
Conclusion 2: The force of gravity on MetOp A will be larger than the force of gravity on Iridium.
Conclusion 3: Inmarsat-3 will have the largest force due to gravity.

Enquiry:
Journey to the Moon

The Apollo Program was designed to land human beings on the Moon. Between 1969 and 1972 six Apollo missions landed on the Moon.

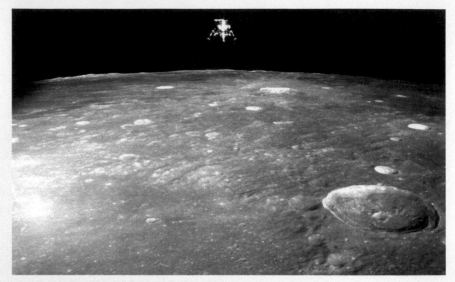

▲ The Apollo 12 Lunar Lander photographed from the orbiting command module by astronaut Richard R. Gordon Jr. The photo was taken on 19 November 1969

NASA engineers and scientists used physics to work out how fast the rockets carrying the space ships had to travel to escape the pull of Earth's gravity. They also chose the best flight paths for the Apollo space ships as they travelled between the Earth and the Moon.

The diagram shows a simplified journey to the Moon followed by the lunar missions.

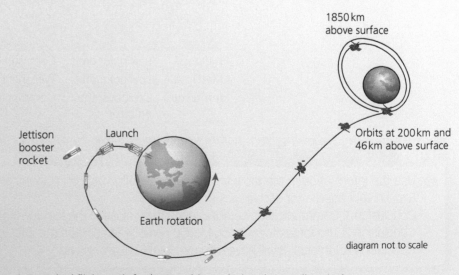

1850 km above surface

Jettison booster rocket

Launch

Orbits at 200 km and 46 km above surface

Earth rotation

diagram not to scale

▲ A typical flight path for lunar orbiters during the Apollo missions

The Apollo spaceships did not have instruments to measure gravity as they flew to the Moon. We can calculate the value of gravitational field strength, g, that the astronauts would have felt.

1 At 190 km above the Earth's surface, gravitational field strength is 9.2 N/kg. Wearing a space suit, an astronaut has a mass of about 160 kg.

a) Calculate the astronaut's weight at 190 km above the Earth.

As the astronauts travelled away from the surface of the Earth, the gravitational field strength on the spaceship decreased. However, near the end of their journey the gravitational field strength on the spaceship started to increase again.

b) Explain why the gravitational field strength on the spaceship changed.

c) Describe how the astronauts' weight changed as they travelled from orbit around Earth to orbit around the Moon.

★ **You can find out more about the Apollo Space Program at: http://www. nasa.gov/mission_pages/ apollo**

Anti-gravity

In the film *Back to the Future*, hoverboards use anti-gravity to move around. Imagine if an anti-gravity hoverboard was invented.

▲ How would life change if we had anti-gravity hoverboards?

2 Write about the costs and benefits of anti-gravity hoverboards. In your answer you should consider:
- which groups of people could benefit or be harmed by the invention of hoverboards
- how could they benefit or be harmed
- if there are financial or environmental issues involved.

Electromagnetism

Learning objectives

3 Voltage and resistance

In this chapter you will learn...

Knowledge

- that voltage is an electrical push from the battery, or the amount of energy per unit of charge transferred through the electrical pathway
- that in a series circuit voltage is shared between each component, while in a parallel circuit voltage is the same across each loop
- that components with resistance reduce the current flowing and shift energy to the surroundings
- the definitions of the terms potential difference (voltage), resistance, electrical conductor and electrical insulator

Application

- how to draw a circuit diagram to show how voltage can be measured in a simple circuit
- how to explain how voltage and resistance affect the way components work, referring to energy
- to use the ratio of voltage across a component to current through it to determine the resistance
- how to explain why part of a circuit has higher resistance
- how to calculate resistance using the formula: resistance (Ω) = potential difference (V)/current (A)

Extension

- how to predict the effect of changing the rating of a battery or a bulb on other components in a series or parallel circuit
- how to explain changing voltages in a circuit, using arguments based on energy
- how to explain safety risks, referring to voltage, resistance and current

4 Current

In this chapter you will learn...

Knowledge

- that current is a movement of electrons and that it is the same everywhere in a series circuit
- how current divides between loops in a parallel circuit, combines when loops meet, lights up bulbs and makes components work
- how the electric field affects other charged objects, causing them to be attracted or repelled
- that field strength decreases with distance
- that two similarly charged objects repel and two differently charged objects attract
- the definitions of the terms negatively charged, positively charged, electrons, charged up, electrostatic force, current, in series, in parallel and field

Application

- how to describe how current changes in series and parallel circuits when components are changed
- how to build series and parallel circuits from diagrams
- how to draw circuit diagrams from the real thing
- how to describe what happens when charged objects are placed near to each other or touching
- how to draw a diagram to show how an object charged positively or negatively became charged up

Extension

- how to compare the advantages of series and parallel circuits for particular uses
- how to evaluate a model of current as electrons moving from the negative terminal to the positive terminal of a battery, through the circuit
- how to suggest ways to reduce the risk of getting electrostatic shocks

Voltage and resistance

Your knowledge objectives:

In this chapter you will learn:

- that voltage is an electrical push from the battery, or the amount of energy per unit of charge transferred through the electrical pathway.
- that in a series circuit, voltage is shared between each component while in a parallel circuit voltage is the same across each loop.
- that components with resistance reduce the current flowing and shift energy to the surroundings.
- the definitions of the terms: potential difference (voltage), resistance, electrical conductor, electrical insulator.

See page 29 for the full learning objectives.

» Transition: Making things work

It would probably be difficult to count the electrical devices a person in Britain uses each day.

Each device seems very different because when switched on, something different happens. It might move or light up. It might make a noise or warm something up. Often it will be a combination of these. But these changes are all caused by what we call a **potential difference** or **voltage**.

Some devices only work when plugged in. Others have a battery which must be replaced or 'recharged'.

> **Key word**
>
> **Potential difference** is the amount of energy shifted from the cell to each unit charge, or from each unit charge to a component, in volts (V). Sometimes called voltage.

Building circuits

To understand potential difference, start with a simple circuit: a cell connected by wires to a lamp (light bulb).

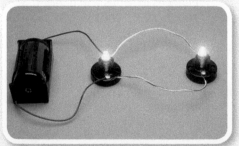 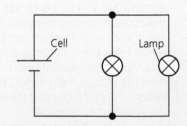

▲ The circuit diagram makes sure the components are connected in the right way

The potential difference across other components, such as a lamp, can also be measured. If there is just one lamp, the measurement will be the same as the cell. If there are two lamps, the measurements will add up to be the same as the cell. If potential difference is higher, more energy is transferred.

> **Key fact**
>
> → The scientific word 'battery' means more than one cell connected together to increase the potential difference. Often 'battery' is used in everyday life when a scientist would say 'cell'.

Key words

Charges are tiny particles in wires and components.

Resistance measures how hard it is to push charges through a material or component. It is measured in ohms (Ω).

An **electrical conductor** is very easy for charges to move through.

An **electrical insulator** is very hard for charges to move through.

Key fact

→ We can measure the potential difference of the cell by using a voltmeter. It is often easiest to connect the voltmeter last.

A potential difference *pushes* **charges** out of one end of the cell and *pulls* them in at the other end. The charges can only move if there is a loop, called a complete circuit.

Common error

The cell doesn't *make* charges – the wire has plenty! The cell makes them move, like a person's heart pumping blood around their body.

It is easy to push the charges through some materials. They are described as **electrical conductors** and have a low **resistance**. Other materials have a high resistance. This means it is hard to push the charges through them. If the resistance is high enough, it is almost impossible to move the charges and we call the material an **electrical insulator**.

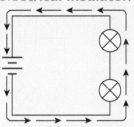

▲ All the charges move at the same time

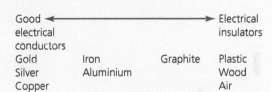

Good electrical conductors			Electrical insulators
Gold	Iron	Graphite	Plastic
Silver	Aluminium		Wood
Copper			Air

▲ Some materials are better conductors than others

Worked example

A cell provides 2.4 V. There are two bulbs in the circuit. We measure the potential difference across the first as 1.3 V. What is the measurement for the second?

2.4 – 1.3 = 1.1 V

The potential difference across the second bulb is 1.1 V, because the total must be the same as the cell provides.

Know >

1 Give the unit for:

a) potential difference

b) resistance.

2 What is pushed around the circuit by the potential difference?

3 Does a good conductor have a high or low resistance?

Apply >>

4 Alex tests each of these solids to see if they complete a circuit. Predict whether they will be conductors or insulators.

a) Wooden stick

b) Steel paperclip

c) Red ribbon

d) Glass rod

e) Plastic ruler

f) Silver hairclip

5 Work out the missing potential difference in each row.

Cell	Bulb 1	Bulb 2
3 V	1.5 V	
4.5 V		1.5 V
	0.7 V	0.8 V

6 A circuit has been left connected overnight and the bulb won't light up any more. Explain what is wrong with *each* idea.

It's not working because the charges have run out.

There's more resistance now so that stops it working.

➤➤ Core: Potential differences

The two sides of a cell have a difference in potential between them. If they are connected together by wires or components, this 'pushes' charges in a loop around the **circuit**. Different cells contain different substances, so the push is a different size. A voltmeter measures the potential difference between any two points in a circuit.

If there are two cells connected, the pushing forces can add up or cancel out. This is why care must be taken when making connections in a circuit. Diagrams help to show how components are linked together. Some other components need to be connected in a particular direction too, such as LEDs.

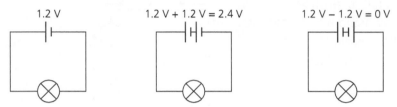

▲ The direction of the symbol shows which way to connect the cell

As we have already seen, more than one cell connected together is called a battery. Lots of cells together mean a bigger potential difference across the components in the circuit. If the other components are the same, this means a bigger effect on the components. This could mean faster heating or a brighter light from a bulb.

Common error

If your circuit doesn't seem to work, take out the voltmeter and try again. Voltmeters must always be extra, not part of the main loop.

▲ The potential difference supplied by a bolt of lightning can be a hundred million volts

In most circuits, potential difference is supplied from cells or the mains electricity supply. In a simple loop this potential difference is shared between the components. If the potential difference across each component is measured, they add up to the same as is supplied. If there are more components, they each get a smaller share. The shares are only equal if the components are identical.

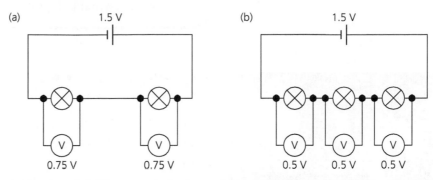

▲ The potential differences *in* the loop add up to the potential difference *across* the loop

Know >

1 Give two examples of sources of potential difference.

2 What unit is used to measure potential difference?

Apply >>

3 Claire is connecting 1.2 V rechargeable cells to see what happens. She tests the overall potential difference.

a) What device does she need for the measurements?

b) What will the reading be for **three** of the cells in the same direction?

c) Claire adds another cell and the reading drops to 2.4 V. Explain why this has happened.

4 A 9 V battery is taken apart. There are six separate identical cells. What potential difference does each of them supply?

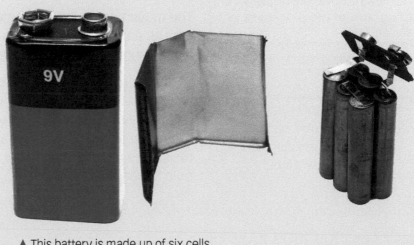

▲ This battery is made up of six cells

5 A security camera needs a supply of 24V. How many 1.5V AA cells would you need?

Extend ≫

6 Where does the word 'battery' come from? Why is it used for a set of cells working together?

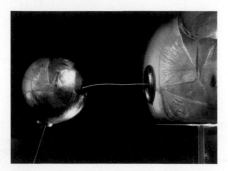

▲ Even an air gap conducts if the potential difference is high. We can see, hear and feel a spark

Common error

Students often mix up the symbol and unit for current. The symbol for current is *I*, which originally stood for Intensity. The unit is the ampere or amp, shortened to A.

Key words

The flow or movement of charge is called **current** and is measured in amperes (A).

Key fact

→ A component which follows Ohm's Law (so has a fixed resistance) is described as *ohmic*.

≫ Core: Resistance

Charges are pushed around the loop by the potential difference. Depending on the material, it can be easy or hard for the charges to move.

A simple model is that most metals are electrical conductors which let charges move easily, and non-metals are electrical insulators which don't. The real situation is a little more complicated. In fact, if the potential difference is high enough it can make charges move through pretty much anything, even air.

The real world is not divided into perfect conductors and perfect insulators. Some materials are harder for charges to travel through than others; scientists like to measure with numbers instead of just describing with words. The quantity is called resistance *(R)* because the material *resists* the movement of charges. It is measured in ohms, shortened to Ω (the Greek letter omega).

For some components and materials the resistance has a fixed value and can be worked out. They follow a rule called Ohm's Law which means that if double the potential difference is *across* them, twice as much charge moves *through* them each second.

The movement of these charges is called **current**. It is measured with an ammeter. Saying 'amp-meter' five times quickly is a good way to show why the spelling changed.

If the potential difference across a component (measured with a voltmeter, in volts) and the current through it (measured with an ammeter, in amps) are known, this relationship calculates the resistance (in ohms). Not all components will have the same resistance in every circuit. For example, the resistance of a filament bulb increases when it heats up.

In a simple loop, the overall resistance in the circuit is the same as the total of all the different resistances added together. Components called resistors are used so circuits work properly; they often heat up when current goes through them.

Key fact

→ resistance (Ω) = potential difference (V) / current (A)

$R = \dfrac{V}{I}$

★ **Other forms of this equation are explained in Chapter 4.**

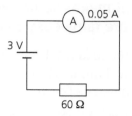 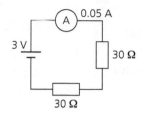

▲ Two 30 Ω resistors in a loop behave the same as one 60 Ω resistor

Worked example

There is a current of 0.2 A through a component when a potential difference of 4 V is across it. What is the resistance of the component?

$R = V/I$

$\quad = 4 / 0.2$

$\quad = 20\,\Omega$

Know >

1 For each pair of materials, state which is the better conductor:

a) wood, silver

d) graphite, silver

b) copper, iron

e) copper, wood.

c) plastic, graphite

Apply >>

2 Copper wire has a low resistance, but silver is an even better conductor. Why don't we often use silver in electrical cables? What other characteristics of copper are important?

3 Hannah's hand slips and connects two wires with a potential difference of 5000 V. A current of 0.005 A flows. What is her resistance?

4 Work out the missing value in each line of the table.

Resistance of component X (Ω)	Resistance of component Y (Ω)	Resistance of component Z (Ω)	Total resistance (Ω)
450	700	450	
	350	350	1200
120	240		800

Extend >>>

5 Graham is asked to work out the resistance of a mystery component. Draw the circuit he would need to build, including any meters.

» Extend: Series and parallel

Everything – wires, bulbs, tables, air, humans – is full of charges. What makes something a good conductor is that the charges are fairly easy to move in a particular direction.

Potential difference is what makes them move. Resistance is a way to measure how hard it is to move them. The gradual movement of charges is what we call current. As the charges move through wires and components they do electrical work. This might increase the temperature, cause movement or any of the other effects useful in electrical circuits. This is why the idea of energy is used in the definition; a larger potential difference across a component means each charge does more work as it moves through it.

Series and parallel

A simple loop can be described by some simple maths. Adding up all the resistances gives the total resistance. Adding up the potential differences across each component gives the supplied potential difference. But what about more complicated circuits?

A simple loop is often called a **series** circuit. There is only one way for the charges to move. However, components don't have to be one after the other, in sequence. They can be in different branches, making more than one way for charges to go. These branches are called **parallel**.

The bulbs in a parallel circuit are brighter than those in a series circuit. Measuring the potential difference across them helps to explain why.

In this example – with nothing else in the main loop – the potential difference across each branch is the same as the cell provides. There could be three, four or ten parallel branches and the result would be the same.

Each branch is independent of the others. Add a resistor, add a bulb, or break the connection completely: the other bulbs are not affected. This means they can be controlled individually. In a series circuit, any change will affect the whole loop, and maybe even break it, for example when a bulb is damaged. This is a problem because there needs to be a complete circuit for charges to move.

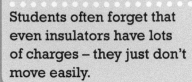

Common error

Students often forget that even insulators have lots of charges – they just don't move easily.

Key words

Components are in **series** if they are in the same loop.

Components on different loops are said to be **parallel** to each other.

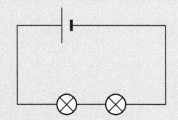

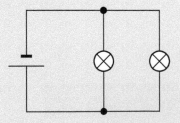

▲ In a circuit, the connections matter as well as the components

Key fact

→ When a switch is *closed* it means the circuit is complete, so charges can move.

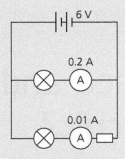

6 V

0.2 A

0.01 A

▲ The second loop
has a resistor as
well as a lamp

To find the resistance of each branch in the diagram on the left we
need to use the equation:

$$R = \frac{V}{I} \qquad R = \frac{V}{I}$$

$$= \frac{6}{0.2} \qquad = \frac{6}{0.01}$$

$$= 30\,\Omega \qquad = 600\,\Omega$$

The higher current means the first bulb would probably be much
brighter.

Tasks

1. For the example above, find the value of the resistor assuming the
   bulbs are identical.
2. By looking for complete loops, state which of the bulbs (X, Y
   and Z) would light in the circuit below for each combination of
   switches.

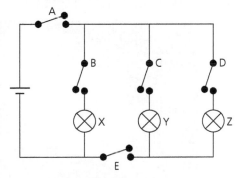

▲ Components A, B, C, D and E are switches

   a) Switches A, B and E closed
   b) Switches A, C and D closed
   c) Switches B, C and D closed
   d) Switches A, D and E closed

3. For the circuit in the diagram above, find the resistance of each
   identical bulb when a 9 V cell is used and the current in each
   branch is 0.45 A. Can you predict what reading would be found if
   the current was measured next to the cell?
4. Draw a circuit with three bulbs and a switch. One bulb should be
   on all the time but the other two must be controlled together by
   the switch.
5. Returning to the diagram **on page 36** predict how long a cell with ten
   loops would last compared to the version with just one loop. Why?

Enquiry:
How do the potential difference and current change in a series circuit?

★ See http://courseweb.
stthomas.edu/apthomas/
SquishyCircuits for more
information and recipes.

Small children like playing with Play-Doh. A version can be made in any kitchen; flour and water with added salt, colouring and a little oil to make it smooth. It can have glitter in it, too. Non-toxic, cheap and fun, it's popular with parents and play groups. And now with science teachers.

It turns out that if made with extra salt and perhaps some lemon juice, the dough is a reasonable electrical conductor. It's not as good as most metals, but it can replace copper wire in a circuit well enough that an LED lights up.

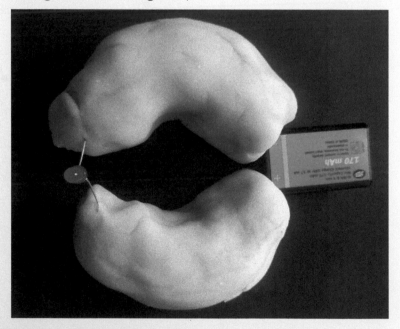

▲ Look, no wires!

❶ a) Draw a circuit diagram for an experiment that will measure the current through the LED. (There is no special symbol for the conducting dough; just draw it as wire.)

Number of cells	Current (mA)
1	0.42
2	0.83
3	1.24
4	1.68
5	2.10

b) Describe the relationship between the number of cells and the current.

c) There is no reading for zero cells. What would the current be?

d) Predict the current flowing if six cells are used.

❷ A student decides to use a block of the dough as a resistor in a circuit. To calculate the resistance of the dough they will need to measure the potential difference across it.

a) What device will be needed to do this?

b) Using a resistor symbol for the dough, draw a circuit diagram for this experiment.

Potential difference (V)	Current 1 (mA)	Current 2 (mA)	Current 3 (mA)	Current (mA) mean to 2 decimal places
1.4	0.42	0.42	0.41	0.42
2.8	0.83	0.84	0.82	
4.1	1.24	1.23	1.23	1.23
5.6	1.68	1.70	1.69	1.69
6.9	2.10	2.12	2.09	2.10

c) Calculate the mean missing from the table.

d) Draw a graph of the mean current (mA) against the potential difference (V).

e) What is the **gradient** of the graph? Include the units.

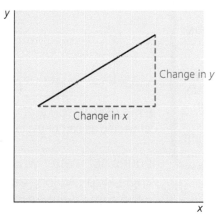

▲ Your graph will not be the same as this diagram, but drawing a triangle may help to find the vertical and horizontal change

f) Explain how this value is linked to the resistance of the dough sample.

» Transition: Potential difference or current?

In electrical circuits, potential difference is what makes things happen; it is what pushes the charges around the loop. All of the charges move at the same time. Some materials are easier for the charges to move through than others. This is measured as a quantity called resistance.

The movement of charges is called current. Current is measured with an ammeter, using units of amperes or amps (A). The amount of current controls how bright bulbs are, how fast motors turn and the volume of a loudspeaker.

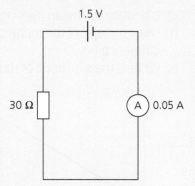

▲ The ammeter is always connected as part of the loop

Common error

Students often think of current as the speed of the moving charges. A much better description is that the current tells us how many charges go through the ammeter each second.

It does not matter where the ammeter is put in a series circuit (a single loop). As long as the connections do not change, the current is the same everywhere in the loop so the reading is the same.

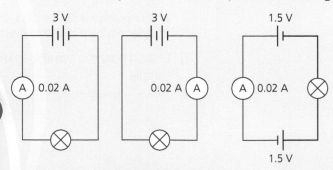

▲ The current is the same everywhere in a series circuit

Two circuits might have the same current even though the potential difference from the cell is very different. It all depends on how many other components are in the circuit and their total resistance.

Common error

Just because the current is the same does not mean the bulbs will all be the same brightness. The same current makes some bulbs bright and others dim.

Key facts

→ If the potential difference is increased, the current will increase (be higher).

→ If the resistance is increased, the current will decrease (be lower).

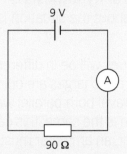

9 V

90 Ω

▲ What will the reading on the ammeter be?

If talking about the potential difference, scientists mean how hard the 'push' is on the charges, or how much work the charges can do as they move through a component. Current means how many charges are moving around the loop.

Calculations

Rearranging Ohm's Law calculates the current in a series circuit using the potential difference and resistance.

current (A) = potential difference (V)/resistance (Ω)

$$I = V/R$$

$$I = V/R$$

$$= 9/90$$

$$= 0.1A$$

There are simple rules to remember for any loop.

Worked example

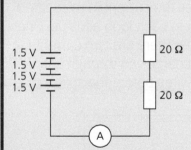

1.5 V
1.5 V
1.5 V
1.5 V

20 Ω

20 Ω

A

What will the reading on the ammeter be?

First, the overall values for potential difference and resistance must be worked out.

Potential difference = 4 × 1.5 V = 6 V

Resistance = 2 × 20 Ω = 40 Ω

$$I = V/R$$

$$= 6/40$$

$$= 0.15 A$$

Know >

1 Draw circuit symbols for the following components.

a) Cell

b) Lamp

c) Resistor

d) Ammeter

e) Voltmeter

f) Switch

2 Will a copper wire have a high or low resistance?

3 What does a voltmeter measure?

4 Give the unit and abbreviation for the following.

a) Potential difference

b) Resistance

c) Current

Apply >>

5 Find the current for each set of values. Remember to show your working.

a) 230V, 4600 Ω

b) 24V, 960 Ω

c) 3V, three 200 Ω components in series.

6 What will the new current be if a second identical cell is added to circuit at the start of this chapter?

7 Draw a circuit diagram to measure potential difference across a mystery component as well as the current through it. Explain how this will allow you to calculate its resistance.

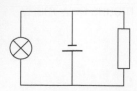

▲ In one loop the lamp is dimly lit, in the other the resistor gets warm

Key word

A **hypothesis** is a prediction which can be tested by experiments or observations.

» Core: Multiple loops

In Chapter 3 we saw that if there is more than one loop, the potential difference across each loop is the same. The effect might be very different – it depends on what is in that part of the circuit!

Scientists predict what they think will happen before doing an experiment. This helps them tell whether their idea about how the universe works is correct. If the scientific idea or **hypothesis** is not right, it means the way they are thinking about the situation is missing something out.

For current, scientists can predict what the values will be in different parts of a series circuit by thinking about what the charges are doing. When the circuit branches, the charges can't travel both parallel ways at the same time. The push across each branch is the same; this can be checked with a voltmeter. To find the current, an ammeter must be connected in each part of the circuit.

Looking at the diagram below, it should be clear that ammeters Y and Z will not have the same reading. They have the same potential difference across them (6V) but different resistances. In fact ammeter Y reads 0.3A and ammeter Z reads 0.1A.

Ammeter X has another reading. It gives the current in the shared parts of the circuit as 0.4A. The current in each loop adds up. One way to understand this is to imagine that there are two currents going through the shared parts. Some charges go through each loop, but they must all go through the wires before the split.

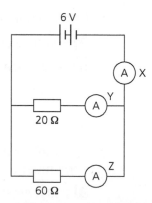

▲ What will each ammeter read?

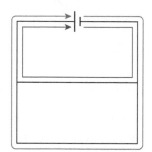

▲ Some parts of the parallel circuit have more current than others

Higher currents cause more heating of wires. They mean brighter bulbs, louder buzzers and faster motors. Even if each parallel branch in a circuit has a low current separately, they add up and can cause problems. This is why it can be dangerous to have many electrical devices plugged into one mains socket. Parallel and series circuits are useful in different situations.

▲ The high total current could start a fire

Worked example

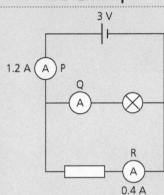

3 V

1.2 A (A) P

Q

R

0.4 A

What will the reading of ammeter Q be?

I = 1.2 − 0.4
 = 0.8 A

Know

1 State a rule for the potential difference across parallel loops.

2 What happens to a wire carrying a high current? Why is this possibly dangerous?

Apply »

3 A third loop is added to the circuit above, and the ammeter in it reads 0.6 A. What will the new reading be for ammeter P?

4 Three identical lamps are connected in parallel. The current in the main circuit is 2.7 A. What is the current in each loop?

Extend »»

5 Work out the resistance of each of the two loops in the diagram above as it is drawn, and for the whole circuit. What do you notice?

6 For a scientist, what is the difference between a hypothesis and a theory?

» Core: Statics

So far, all the examples have used moving charges. That's how current works, after all. But what if the charges can't move?

Everything contains charged particles. Usually, the positive and negative charges are in balance. In most solid materials the negative charges, called **electrons**, are the ones which move around. They can be trapped in one place, or moved from one object to another by friction. This means the charges aren't balanced any more and because they can't move easily, the effect is called static charge. One object will be **positively charged**, the other **negatively charged**.

Key words

Electrons are tiny, **negatively charged** particles that are part of atoms.

An object which has lost electrons will be **positively charged**. An object which has gained electrons will be negatively charged.

▲ Each hair has the same charge so they are pushed apart

Two negatively charged objects will experience a force pushing them apart. This is called repulsion. The same thing happens with two positively charged objects. Objects with opposite charge – one positive and one negative – will be pulled together instead, which is called attraction.

Any charged object will have an **electrostatic field** which is strongest close to the object.

Key word

In an **electrostatic field** any charged object or particle experiences a force.

The separation of positive and negative charge makes a potential difference. This will cause a current to flow, just like a cell or battery, *if* there is a connection. Instead of a complete loop, static charge usually flows when the charge can get to the Earth. This current is called grounding or earthing and is usually brief, but sometimes dramatic.

Key fact

→ These electrostatic forces, like magnetism and gravitational attraction, are non-contact forces.

Common error

Students often mix up electrostatics with magnetism. The rules of attraction and repulsion are the same, but the causes are different. See Pupil's Book 2, Chapter 4.

Sparks

▲ A potential difference of 30kV can cause a big spark

Because the potential difference is much higher than seen in most circuits, even air can briefly conduct. This causes a spark, which may be seen and heard – or even felt! Lightning is effectively a very big spark. Insulators rubbing together on hot, dry days can cause the negative charges to get trapped and out of balance. Unless the charge can travel through a conducting grounding cable, the sparks can start fires or damage electronic equipment.

Know >

1 What are the small negative charges called?

2 What force is opposite to attraction?

Apply >>

3 Two objects, both positively charged, are moved closer together. What will happen to the force between them?

4 A racing car driver's fireproof suit rubs against the seat, losing negatively charged particles.

   a) Is the suit now positively or negatively charged?

   b) What will happen when the driver gets out of the car, touching the ground?

5 Suggest why a petrol tanker is connected to the ground with a copper cable before it is filled up with fuel.

Extend >>>

6 Why is lightning more likely to hit tall objects? What do we do to protect skyscrapers?

» Extend: Models

Something which often confuses students is the movement in electrical circuits. People can't see what is moving, so models are needed to understand what is happening. A scientific model is just a way to think about what is happening in a situation so useful predictions can be made.

Students often think of charges in a circuit as leaving the cell, carrying something to the components to make them work, and returning to the cell to be refilled. They imagine this happening really quickly, because when a switch is closed a lamp lights up immediately.

A better model is that the moving charges are all in a row. As some are pushed out of one end of the cell, the same number are pulled in at the other end. Because all of the charges move at the same time, like the links in a bicycle chain, there is no delay. As soon as the potential difference makes *one* move, they *all* move.

This can be modelled in real life with students in a circle, holding a rope loop. One student pulls the rope sideways. The whole loop moves at once, and every student feels it move through their fingers. If one student in the loop holds it tighter, this makes it harder to move.

▲ Comparing models makes them easier to understand

Feature of the model	Representing
Pulling the rope	Potential difference
Movement of rope	Current
Gripping tighter	Increasing resistance
Material of the rope	Electrons in the wire

Mistakes

"In science it often happens that scientists say, 'You know that's a really good argument; my position is mistaken,' and then they would actually change their minds and you never hear that old view from them again. They really do it. It doesn't happen as often as it should, because scientists are human and change is sometimes painful. But it happens every day."

Carl Sagan

One of the first scientists to describe electricity, Benjamin Franklin, made one simple mistake. He assumed that the moving particles were *positively* charged. In fact, the electrons have a negative charge.

All the rules are the same, all the maths still works. But it means current can be imagined in two ways. 'Conventional current' is described as flowing from the positive to the negative side of the cell. Electron current moves in the opposite direction. It doesn't make any difference, but some students find it helps them understand the situation better to know the reason.

Tasks

1. The rate at which charges move is called current. Where do the particles come from?
2. Megan says "I think it will rain soon because of the clouds." Is this a scientific prediction?
3. What other features of the rope model have links to what is really happening in the circuit?
4. Why do you think scientists don't give up on conventional current and use the more accurate description all the time?
5. Benjamin Franklin is famous for more than just science. Research the important things in his life and write about how scientific skills – making careful observations, collecting evidence, looking for patterns, communicating clearly – may have been useful.

▲ Carl Sagan, an American astrophysicist

Energy

Learning objectives

5 Energy costs

In this chapter you will learn...

Knowledge

- about the information given on food labels
- the definitions of renewable and non-renewable in terms of energy resources
- some examples of renewable and non-renewable energy resources

Application

- how to calculate the cost of home energy usage, using the formula: cost = power (kW) × time (hours) × price (per kWh)
- how to compare the energy usage and cost of running different home devices
- how to explain the advantages and disadvantages of different energy resources
- how to represent the energy transfers from a renewable or non-renewable resource to an electrical device in the home

Extension

- to evaluate the social, economic and environmental consequences of using a resource to generate electricity, from data
- how to suggest actions that a government or communities could take In response to rising energy demand
- how to suggest ways to reduce costs, by examining data on a home energy bill.

6 Energy transfer

In this chapter you will learn...

Knowledge

- how the energy of an object depends on its speed, temperature, height or whether it is stretched or compressed
- some examples of energy stores: thermal, chemical, kinetic, gravitational potential, and elastic
- about the changes in a system using energy stores at the beginning and end

Application

- how to show how energy is transferred between energy stores in a range of real-life examples
- how to calculate the useful energy and the amount dissipated, given values of input and output energy
- how to explain how energy is dissipated in a range of situations

Extension

- how to explain why processes such as swinging pendulums or bouncing balls cannot go on forever, in terms of energy
- how to evaluate analogies and explanations for the transfer of energy

5 Energy costs

» Transition: Energy

Your knowledge objectives:
In this chapter you will learn:
- about the information given on food labels
- the definitions of renewable and non-renewable in terms of energy resources
- some examples of renewable and non-renewable energy resources

See page 49 for the full learning objectives.

Common error

Sometimes people use energy to explain why things happen. However, things happen because of forces, and you should always describe changes in motion by using the forces and physical processes, not energy.

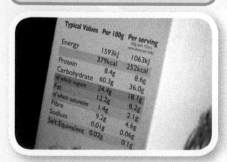

▲ Food labels let you know what you're eating

Key fact

➡ Food labels list the energy content of food in kilojoules (kJ).

It's difficult to describe what energy is, in scientific terms. However, energy calculations let us answer useful questions such as:

- **How much pasta do I need to eat to run a marathon?**
- **How much electricity do you need each week to charge up your mobile phone?**
- **How many tonnes of plankton do blue whales need to eat to live?**
- **If I mix a hot liquid and a cold liquid, what will the final temperature be?**

Energy from food

Look on the back of most food packages, and you will see a **food label**. It provides information about the food you're eating and what it contains.

On a food label you can see different values given for energy: kJ and kcal. Calories (written as kcal) are an older unit of measurement, and kilojoules (written as kJ) are the unit we use in science. One calorie (kcal) is the same as 4.2 kJ.

The energy content of food comes from three different food groups:

- fats
- proteins
- carbohydrates (starches or sugars).

The table shows the standard values used for the energy content of 1 g of different food groups.

Food group	Energy in 1 g (kJ)
Fats, e.g. butter, oil	37
Proteins, e.g meat, seeds	17
Carbohydrates – starches and sugars, e.g. bread, fruit	16

When you eat, your body breaks down the food into simpler molecules. Proteins are used for growth and repair. Carbohydrates and fats are used in respiration to provide energy. Our bodies are

▲ One Jaffa cake contains 190,000 J of energy. One joule is a very small amount of energy so we use kilojoules to measure energy from food

★ **You can read more about food groups, digestion and respiration in Pupil's Book 2, Chapters 16 and 17.**

also very good at storing fat. If we eat more fat than we need for energy, then it is stored in fat cells for later use.

The table below shows the typical amount of energy transferred during different activities. The actual amount will depend on the person.

Activity	Time (minutes)	Energy (kJ)
Cycling	15	188
Dancing	15	220
Playing netball	15	375
Playing the drums	15	188
Sitting and writing	15	120
Walking	15	143

The energy content of one Jaffa cake is about the same as the amount of energy transferred when cycling for 15 minutes.

Worked example

This diagram shows part of a food label from a yogurt snack. Calculate the energy content of the yogurt snack.

The energy content is equal to the energy content due to fats, carbohydrates and proteins.

6.9 g of fat. Energy content = 6.9 × 37 kJ = 255.3 kJ
13.1 g of carbohydrate. Energy content = 13.1 × 16 kJ = 209.6 kJ

4.2 g of protein. Energy content = 4.2 × 17 kJ = 71.4 kJ
Total energy = (255.3 + 209.6 + 71.4) = 536.3 kJ

Greek style yogurt, granola and apples with pistachio nuts

NUTRITION Typical values per 100 g:

Fat 6.9 g

Carbohydrate 13.1 g

Protein 4.2 g

Salt 0.13 g

Apply >>

1 A bar of chocolate has a mass of 41 g. The food label states that the chocolate contains 12 g of fat. Calculate the energy content from the fat in the chocolate.

2 The food label of a ham sandwich gives the energy content as 861 kJ per 100 g. The sandwich has a mass of 180 g. Calculate the energy content of the sandwich.

3 How much energy would be transferred if you played the drums for 40 minutes? Use the table above to help you.

4 Fruit is a popular snack after exercise. Jo eats 50 g of dried fruit after going for a 30-minute walk.

The picture shows part of the food label on the dried fruit packet.

Compare the energy content of Jo's snack with the energy transferred during her walk.

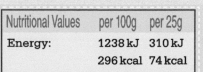

Nutritional Values	per 100g	per 25g
Energy:	1238 kJ	310 kJ
	296 kcal	74 kcal

» Core: Energy resources

Food is an example of an **energy resource**. We can think of food as a type of **fuel**.

Firewood was the earliest fuel to be used by humans over 2 million years ago. It was portable, and easily available. Coal was first used as a fuel around 3 thousand years ago in China. The UK industrial revolution of the 18th century was only possible because of coal.

The table shows the energy content of 1g of different fuels. These fuels provide a convenient store of energy, which can be released through burning.

Fuel	Energy in 1 g (kJ)
Firewood	15.8
Coal	28.8
Oil	12.8
Petrol	13.0
Gas	14.9

▲ A solar panel in use. The Sun is an energy resource

▲ Four examples of fuels (a) coal (b) crude oil (c) petrol (from crude oil) and (d) gas

Coal, oil and gas are all **fossil fuels**. They were formed millions of years ago. Over a very long time, the remains of plants and animals were buried and compressed until coal, oil and gas were formed, as shown in this diagram.

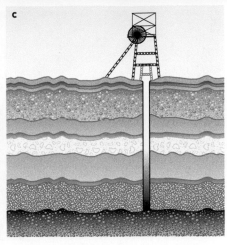

▲ (a) Plants grow using energy from the Sun. Animals eat the plants, and each other. (b) Plants and animals die. Dead material sinks to the bottom of swamps and oceans. It is buried by many layers of sediment. (c) Over millions of years, the temperature and pressure turn the dead material into coal (plants) or oil and gas (animals)

Key word

A **non-renewable** energy resource is one that cannot be replaced and will eventually be used up. Examples are coal, crude oil and natural gas.

The world is burning fossil fuels at a fast rate. Fossil fuels are **non-renewable**. They took so long to form, that we cannot replace them. One day they will run out.

Another example of a non-renewable energy resource is nuclear fuel. One gram of nuclear fuel releases about 80 000 MJ (1 MJ = 1000 kJ). One disadvantage of nuclear fuel is that the radiation released can be dangerous to living organisms.

Common error

Although it's called a fuel, in a nuclear reactor the nuclear material isn't burned like coal or gas.

Key word

An energy resource that can be replaced as it is used, and will not run out is **renewable**. Examples are solar, wind, waves, geothermal and biomass.

Renewable energy resources

Some energy resources can be replaced as they are used. These are **renewable** energy resources. Solar panels are an example of technology that uses sunlight as an energy resource. Solar panels can be placed on buildings to generate electricity.

Biomass is the name for fuel formed from biological matter. Fast growing plants are an example of biomass. Biomass is a renewable energy resource. As biomass is used, we can grow more to replace it. One disadvantage of biomass is that less land might be available for growing food.

Iceland is a country which has many active volcanoes. The rocks deep underground are hot. Geothermal power stations such as the one shown overleaf are used to generate electricity using the hot rocks. Very few countries are suitable for geothermal energy.

▲ Biomass pellets made from waste wood. These can be burned in boilers and power stations

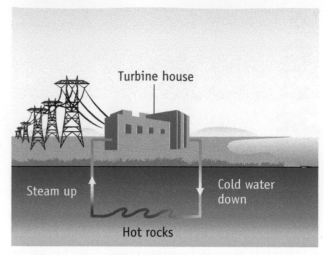

▲ The hot rocks underground heat cold water into steam. The steam turns turbines to generate electricity

Wind and water

Moving water can also be used as a renewable energy resource. Water wheels were used by Ancient Greeks about 4000 years ago. Nowadays, tidal barrages use the daily movement of tides to generate electricity. Ecological concerns mean that not many tidal barrages have been built. Wave power generators are being developed to use the movement of the oceans to generate electricity. However, it is not easy to use waves in this way, so wave generators are rare.

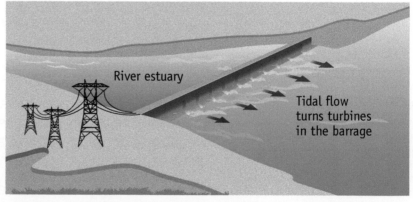

▲ The tidal barrage is a dam across the mouth of a river. Electricity is generated each time the tide moves in and out

▲ Wind turbines can be onshore or offshore. Some people would prefer not to have wind turbines near where they live

Wind has also been used as an energy resource for thousands of years. Modern wind turbines use the movement of air to turn the blades of the turbine. This generates electricity.

Know >

1 Explain why fossil fuels are non-renewable.

2 What is meant by renewable energy resources?

3 Name three examples of renewable energy resources.

Apply >>

4 Each energy resource has advantages and disadvantages. Copy and complete the table to give at least one advantage and one disadvantage for each energy resource.

Energy resource	Advantage	Disadvantage
Non-renewable		
Coal		
Crude oil/petrol		
Renewable		
Solar		
Wind		
Wave		

Extend >>>

5 Compare the energy per gram for foods and for fossil fuels.

6 Suggest why wood is a more useful energy resource to produce than fat.

Enquiry >>>>

A simple experiment to compare the energy content of food is shown in the diagram. As the food burns, the water heats up. The more energy that is transferred, the greater the temperature increase measured on the thermometer.

7 What **variables** would need to be kept constant in this experiment, and why.

8 Rizwan did this experiment using vegetable oil. He compared his value for the energy content of the oil to the standard value. His value was much lower. Explain why.

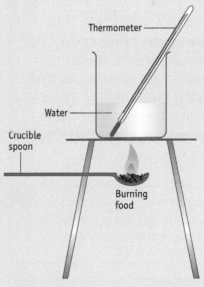

Thermometer

Water

Crucible spoon

Burning food

▲ A simple method of comparing the energy content of different fuels

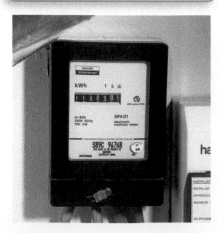

▲ A gas boiler burns gas to heat water. The hot water is then pumped around the house through radiators

Key word

Power is how fast energy is transferred by a device. It is measured in watts.

▲ The meter measures the amount of electricity used in the home

» Core: Energy in the home

Every day we use energy resources in our homes. Many people burn gas in their house for heating and cooking. However, many of the things we use each day are electrical appliances.

▲ Computers and mobile phones all need to use electricity to work

Using electricity

Electricity is generated using energy resources such as coal, wind and the Sun. It is an easy way to transfer energy and do useful things. Our lives would be much less comfortable without domestic electricity.

We have to pay for the amount of energy transferred by electricity. In homes, this is measured using an electricity meter like the one shown in the photo. The amount of energy transferred depends on the appliances that are used, and how long they are used for.

For electrical devices, their **power** is a measure of how much energy they transfer in a certain time. Power of 1 watt is equal to transferring 1 joule of energy in 1 second.

The table shows the power for different devices.

Device	Power (W)
Electric oven	2800
Kettle	3000
Laptop	45
LCD TV	60
Smartphone	5
Toaster	2000
Tumble dryer	2000
Microwave	1000
Games console	200

Key fact

→ Electricity companies charge for the amount of energy that is transferred. This is measured in kilowatt hours (kWh) and is often called a unit. One unit (1 kWh) is equal to 3.6 million joules!

We can calculate how much it would cost to run different devices.

We need to know the power rating of the device, how long it will be used for and the price charged per unit by the electricity companies.

To calculate the cost we use the equation:

$$\text{cost} = \text{power (kW)} \times \text{time (hours)} \times \text{price (per kWh)}$$

Worked example

An electricity company charges 11p per unit of electricity used. Calculate the cost of using a laptop for 8 hours.

Cost = power (kW) × time (hours) × price (per kWh)

Power of laptop = 45W = 45/1000 kW

= 0.045 kW

Mathematical calculation: To calculate cost you must convert power into kW and time into hours.

Cost = 0.045 kW × 8 hours × 11p

Cost = 3.96p

Know >

1 Give two energy resources that can be used to generate electricity.

2 Name the unit of power.

3 What factors does the cost of using an electrical device depend on?

Apply >>

For the following questions assume that the cost of electricity is 11p per unit.

4 A baked potato takes 1.5 hours to cook in an electric oven. Calculate the cost of cooking the potato.

5 It takes 4 minutes to toast two slices of bread in a toaster. Calculate how much this costs.

6 A tumble dryer manufacturer claims that it costs 30p to dry each load of laundry. How long does it take to dry the laundry?

Extend >>>

7 Look at the power ratings of the different electrical devices in the table on the page opposite. Suggest why some of the devices have a much higher power rating than others.

Enquiry >>>>

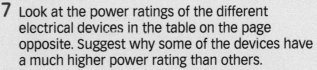

▲ An example of an electricity bill

Electricity bills are made up of two parts: the cost of the units of electricity used and a standing charge. The standing charge is a fixed fee you pay every day even if you don't use any electricity that day.

The table shows the costs for three different electricity plans.

	Cost per unit (p)	Daily standing charge (p)
Plan A	11	16
Plan B	14	9
Plan C	17	0

8 Use the data to decide which would be the best electricity plan for the following people:

a) Julia has an electricity supply for her garden shed. She uses 200 units per year in the shed.

b) Benoit lives in a one bedroom flat. He uses 1900 units per year.

c) A group of students share a large house. They use 4100 units per year.

» Extend: Reducing electricity use

All the energy resources used to generate electricity have advantages and disadvantages. Fossil fuels release carbon dioxide which contributes to global warming. Renewable energy generation technology does not release carbon dioxide but the technology is still developing.

Many governments, including the UK, have committed to reducing the amount of carbon dioxide their countries produce. One way to do that would be to increase the proportion of electricity generated from renewable energy resources. Another way would be to reduce the amount of electricity that is used, so that less needs to be generated in the first place.

Some ways to reduce the amount of electricity used are:

- encourage people and businesses to use less electricity
- make appliances that do the same job using less electricity (more efficient)
- make sure buildings, such as houses and offices, are well insulated.

> You can find out more about climate change in Pupil's Book 2, Chapter 13.

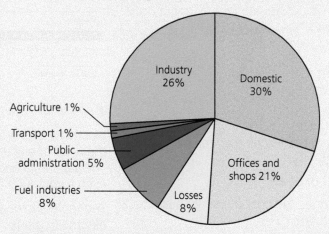

▲ This chart shows the proportion of electricity used by different groups in the UK in 2015

Tasks

1. Give another advantage for people and businesses of reducing the amount of electricity they use.
2. Suggest why 'transport' used only 1% of UK electricity in 2015.
3. Give three ways in which you could reduce the amount of electricity you use.
   You might find it useful to refer to the table on page 56 to help you answer this question.
4. By 2020 electricity companies have to offer all households a 'smart meter'. This would allow people to see exactly how much electricity they are using all the time. Why might 'smart meters' help people to reduce the amount of energy that they use?

Alternative energy resources

Biofuel is a type of biomass that can be made from soybean oil.

In the UK, biofuel is added to petrol. This reduces the amount of carbon dioxide emitted by cars that use it.

Brazil is one of the largest growers and exporters of soybeans.

Over 60% of the Amazon rainforest is within Brazil's borders. From the 1960s onwards trees in some areas the Amazon rainforest were cut down. The land was often used for cattle or to grow crops.

The graph shows Brazilian soybean production from 1997 to 2015. This is shown on the left hand axis. The graph also shows the area of trees chopped down each year on the right hand axis.

▲ Soybeans growing in a field. Soybeans are used widely in the food industry

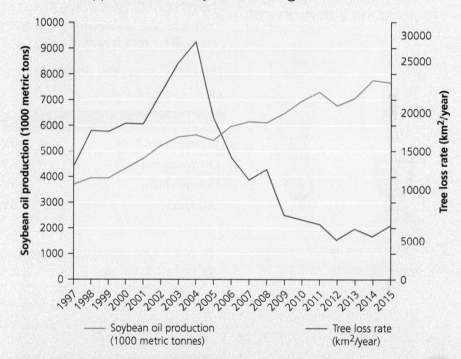

—— Soybean oil production
(1000 metric tonnes)

—— Tree loss rate
(km²/year)

Tasks

5. Explain why biofuel is a renewable energy resource.
6. Describe how soybean oil production has changed between 1997 and 2015.
7. What year had the maximum tree loss rate?
8. Use the graph to estimate the maximum tree loss rate.
9. In 1997 the Amazon rainsforest in Brazil covered an area of 3 576 580 km². By 2015 the area of the Amazon rainforest in Brazil was 3 331 065 km².
   Calculate the percentage reduction in the area of the Amazon rainforest between 1997 and 2015.
10. Environmental campaigners have suggested that some of the land which is cleared of trees is being used to grow soybeans.
    Is there a **correlation** between soybean production and rate of loss of trees in Brazil? Use data from the graph to explain your answer.

Key word

A **correlation** is a relationship between two variables. The dependent variable increases or decreases as the independent variable increases.

Enquiry:
Running costs of light bulbs

▲ An incandescent bulb and energy-saving bulb

Comparing light bulbs

Since 2005, countries around the world have introduced bans on the sale of **incandescent** light bulbs. In the UK, incandescent bulbs were banned from 1 September 2012. After this date, the majority of light bulbs were energy-saving (compact fluorescent), halogen or LED bulbs.

The amount of light a bulb gives out is measured in lumens (lm). If there are two bulbs with the same power rating, an energy-saving bulb will be much brighter than an incandescent bulb.

The table shows the power rating and the typical brightness of a variety of light bulbs.

Brightness (lm)	400+	700+	1300+
Incandescent bulb	40 W	60 W	100 W
Energy-saving bulb	9 W	15 W	20 W
Halogen bulb	28 W	42 W	70 W
LED lamps	6 W	10 W	18 W

❶ Which type of bulb would cost the least to use? Explain your answer.
❷ In the UK, halogen bulbs are also being banned from 2018. Use the data in the table to explain why.
❸ For each type of bulb, calculate how much it would cost to use a 1300+ lumen bulb for 2 hours. Assume that the electricity costs 11p for each unit.
❹ One manufacturer of light bulbs claims: Energy-saving bulbs use 80% less electricity than incandescent bulbs. Is the manufacturer's claim valid? Use data from the table to help you decide.

▲ A photograph of the UK at night taken by Tim Peake from the International Space Station.

Estimating UK savings

Looking at a photograph of the UK at night, it is easy to see that there are a lot of lights on in the houses. But how many? and how much money have we saved by changing incandescent bulbs to energy-saving ones? It's very difficult to calculate accurate answers to these questions. However, we **can** estimate the answers. This is an important skill to use in science when we don't have as much information as we would like.

The following questions will help you to estimate the money saved each year in the UK by changing light bulbs in 2012.

1 How many bulbs in a house?

Count approximately how many lightbulbs you have in your house. Round this figure to the nearest 10.

2 How much money would be saved?

To estimate the cost we need to know how many hours each bulb is used for.

Start by estimating how long a bulb is used each day. Multiply that by the number of days in a year. Then, multiply that by the number of years since the UK ban on incandescent bulbs started in 2012. Finally, multiply that by the number of light bulbs in a house.

Assume that 100 W bulbs were replaced by 20 W energy-saving bulbs. So we can assume that we've used 80 W less per bulb.

Now use cost = power (kW) × time (hours) × price per unit.

What would be a suitable number of significant figures to round this number to?

3 How many houses?

There are about 65 million people living in the UK.

Estimate the size of a typical family.

How many houses would there need to be if every family had their own house?

Round this figure to the nearest million.

4 How much have we saved?

Finally multiply the number of houses in the UK by the amount of money saved per house.

Your figure is an estimate. That is not the same as a guess. You have made reasonable assumptions which have allowed you to end up with an answer to the question with limited information.

★ **You could calculate an average for the number of lightbulbs in a house by pooling data with other people in your class.**

61

6 Energy transfers

» Transition: Energystores

Your knowledge objectives:

In this chapter you will learn:

- how the energy of an object depends on its speed, temperature, height or whether it is stretched or compressed
- some examples of energy stores: thermal, chemical, kinetic, gravitational potential, and elastic
- about the changes in a system using energy stores at the beginning and end

See page 49 for the full learning objectives.

Key words

A **system** is an object, or a group of objects, that interact. We can choose what we include in the system.

A **gravitational store** is filled when an object is raised. This is also known as gravitational potential energy.

A **kinetic store** of energy is filled when an object speeds up. This is also known as kinetic energy.

Energy is one of the big ideas of physics. It allows us to work out what might happen in a system. We might want to know:

- **How many wind turbines would we need to replace one coal fired power station?**
- **Which is the best type of physical exercise to do?**
- **How far can we drive in a car on one tank of fuel?**

Energy helps us to find answers to these questions.

We can think of energy as being in different stores. We can track the energy as it shifts (transfers) from store to store in a system. Later on in your science studies you will also be able to calculate the energy in each store.

Pick a book up off the floor and put it on a high shelf. A **gravitational store** of energy is filled. That happens whenever an object is lifted up in a gravitational field. If you drop the book, then the gravitational store of energy is emptied.

▲ A kinetic store is emptied when a car applies its brakes and slows down

A moving object has a **kinetic store** of energy. This is filled when the object speeds up, and emptied when the object slows down.

▲ An elastic store of energy is filled when the archer pulls back the bow string

Elastic stores of energy are filled by compression or stretching materials or objects. This might be objects such as springs or elastic bands, but the amount a surface is squashed by you standing on it can also be an example of an elastic store of energy being filled.

When a fuel is burned in oxygen (combustion), a **chemical store** of energy is emptied. When chemicals react in a battery, a chemical store of energy is emptied.

When an object is heated, its **thermal store** of energy increases. When a hot pie cools down, then its thermal store of energy decreases. All objects have a thermal store of energy.

Later on in your study of physics you will find out about other stores of energy, including magnetic and nuclear stores.

One analogy we can use for energy is money. Your money can be stored in different ways: as coins and notes in your pocket, as data in a bank account, or even as an I.O.U. note. The money doesn't change, even though it is in a different store. Money also doesn't make things happen: but it does let us know what might happen: what we could buy.

Key words

An **elastic store** of energy is filled when a material is stretch or compressed.

A **chemical store** of energy is emptied during chemical reactions.

A **thermal store** of energy is a measure of the energy stored in a substance due to the vibration and motion of particles. It is sometimes just called thermal energy.

Common error

Energy isn't a good way to explain why things happen. To explain why things happen we need to use physical mechanisms (like forces). For example, a ball falls to the floor because of the force of attraction due to gravity on it, not because it has a store of gravitational energy.

Worked example

Nathan is exercising on a trampoline.

(a) What energy store is filled when Nathan is standing on the trampoline? Explain why.

(b) What energy store is filled when Nathan is up in the air?

(a) Elastic store of energy.
    Nathan stretches the surface of the trampoline.

(b) Gravitational store of energy.

Apply »

1 Give an example of a situation in which:

a) a kinetic store is emptied

b) a gravitational store is filled

c) a thermal store is emptied.

Key fact

→ Energy is measured in joules (J). It doesn't matter which store we are using to calculate energy.

» Core: Energy transfers

The thermal energy store of the drink decreases

The thermal energy store of the hands increases

▲ The thermal store of the mug decreases. The thermal stores of the surroundings increases

Think about what happens when you hold a hot drink. The temperature of the drink decreases, but the temperature of the surroundings increases. The surroundings include you holding the drink, but also the air. The thermal energy store of the drink has decreased. Your thermal store of energy and the thermal energy store of the drink has decreased.

When we're thinking about a system, it's useful to know what energy stores are filled at the beginning and what energy stores are filled at the end. Although we could talk about what happens in the middle of the process, it doesn't help us with our calculations.

▲ If we know the amount of energy in the gravitational store at the top of a rollercoaster ride, we can calculate how fast it will be going at the bottom of the ride

On many rollercoaster rides, the rollercoaster cars are pulled up to the top of the ride. Their gravitational store of energy has been increased. At the bottom of the loop, the gravitational store of energy has decreased, but the kinetic store has increased. Energy has been transferred between stores.

Filling and emptying energy stores

★ **You can learn more about work and forces in Pupil's Book 2, Chapter 5.**

In a moving car, when the driver puts on the brakes the car slows down and stops. The brakes apply a force due to friction on the wheels which slows the car down. The temperature of the brakes and the wheels increase.

We can represent the energy stores of the system at a start point (while the car is moving) and an end point (after the car stops moving). At the start the car has a kinetic store of energy. At the end, the brakes and the wheels have a greater thermal energy store. Energy has been transferred by mechanical work and heating.

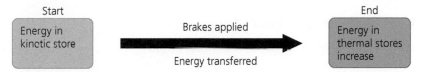

When you boil water in a kettle using a gas hob, the chemical store of energy decreases. The temperature of the water, the kettle and the surroundings increase, so there is an increase in thermal stores.

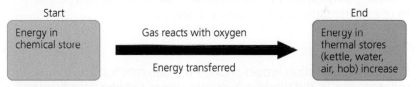

▲ How can we describe the energy stores before and after boiling the kettle?

In both these examples we have chosen a suitable start point and an end point. We can explain what happens to the system between these points using physical mechanisms, but we're only interested in calculating the energy at the start and the end.

Worked example

Draw a simple diagram to show the change in energy stores for a rollercoaster ride between the top of the ride and the bottom of the first drop.

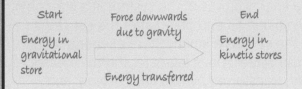

Know >

1 What unit is energy measured in?

2 What is the energy store for a hot cup of tea?

Apply >>

3 Draw a simple diagram to show the change in energy stores for a mug of hot tea as it cools down.

4 A child is playing on a slide. Suggest good start and end points to use to measure the changes in energy stores.

Extend >>>

5 Suggest how friction would change the diagram for a rollercoaster in the worked example above. Explain your answer.

Enquiry >>>>

6 Money is often used as an analogy (or model) for energy and energy transfer. It's important to remember that analogies and models are simplified versions of the thing they are being used to describe. Some aspects of the model will be a good description, but some aspects won't match well. The table gives an example of this for money.

▲ How is a money store like an energy store? How is it different?

How money is a good analogy	How money is a poor analogy
Money and energy can be found in different stores	Energy doesn't exist as an actual thing, but you can have coins and notes of money

Give one more way that money is a good analogy for energy, and one more way that it is a poor analogy.

» Core: Energy dissipation

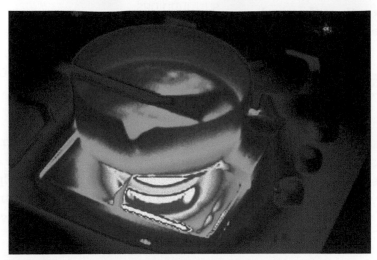

▲ Heating soup with a wooden spoon. The thermogram shows the different temperatures of an object. White is hottest, purple coolest

When you warm food in a pan, it's not just the thermal store of the food that increases. In the thermogram in the photo the temperature scale goes down from white → yellow → red → green → blue → purple. You can see that the flames have a high temperature (white) and that the soup is warming (blue). But you can also see that the pan and the hob both warm up. More than one thermal store has been filled.

Only the energy transfer to warm the soup is useful. The rest of the energy has been **dissipated** in the surroundings by increasing their thermal store of energy.

Energy stores can be concentrated (like the chemical store of a fuel) or spread out (like the thermal store of the hob).

Conservation of energy

When heating the soup, we can calculate the amount of energy in the chemical store at the beginning. We can also calculate the total amount of energy in all of the thermal stores at the end. These two values will be the same. The total amount of energy is conserved.

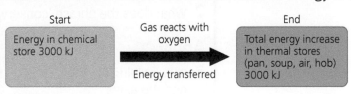

Start | Gas reacts with oxygen | End
Energy in chemical store 3000 kJ | → Energy transferred | Total energy increase in thermal stores (pan, soup, air, hob) 3000 kJ

Wasting energy

Energy is always dissipated by heating, and often this is not useful to us. If energy is transferred we can reduce how much it is wasted by not filling thermal stores of the surroundings.

A car engine uses oil to reduce the friction between the moving parts. This reduces the energy that is wasted.

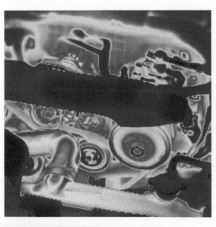

▲ Thermogram of a car engine showing where energy is being dissipated by heating

★ **In Pupil's Book 2, Chapter 6 you will learn more about heating, thermal stores and insulation.**

Key fact

→ In any system, the total amount of energy is conserved when energy is transferred. There is the same amount of energy at the beginning and at the end.

Using insulation in the walls of houses reduces the transfer of energy between the air inside the house, and the house's surroundings.

Although energy is conserved, energy *resources* are not. That is why it is important that we reduce the amount of energy transferred into thermal stores and wasted.

Common error

Sound and light are often described as energy types or stores. However, we don't store energy in either sound waves or light waves. There isn't a helpful calculation we could do. It is better to describe sound and light as energy pathways: how energy is transferred between stores.

Worked example

A pan of soup is heated until it is hot in a pan on a gas hob. The chemical store of energy has reduced by 500 kJ. The thermal store of the soup has increased by 400 kJ.

How much energy has been dissipated?

(a) The total energy in the chemical store = 500 kJ.

Conservation of energy means that the energy of the system before will be equal to the energy of the system afterwards. So the total increase in the thermal stores must also be 500 kJ. The thermal store of the soup has increased by 400 kJ, so other thermal stores must have increased by the remaining amount.

Energy dissipated = 500 kJ – 400 kJ = 100 kJ.

Know >

1 Describe how energy is dissipated when using a kettle to boil water.

2 What does the phrase 'conservation of energy' mean?

Apply »

3 Priti is riding her bike. There is friction between the bike wheels and the axles.

a) Draw a simple energy diagram to show the energy stores that are filling or emptying.

b) How could the energy dissipated at the wheel axles be reduced?

4 Write down one situation where heating by friction is useful.

5 Lizzie is skateboarding. At the top of a half-pipe her gravitational store of energy is 2500 J. Lizzie skates down the half-pipe. On the flat section at the bottom of the half-pipe, Lizzie's kinetic store of energy is 2000 J. How much energy has been dissipated during the skate?

▲ How is energy dissipated on a skateboard at a skatepark?

Extend/Enquiry ⟩⟩⟩

When an incandescent light bulb is turned on, a store of chemical energy at the power station decreases. Energy is transferred electrically to the house. Incandescent bulbs waste a lot of energy by heating up their surroundings.

In a light bulb, the useful energy transfer is by radiation (light). The light does also eventually increase the thermal stores of the bulb's surroundings (or us).

The table shows the energy supplied to each bulb and the energy transferred for three different types of light bulbs.

Type of bulb	Energy supplied each second (J)	Energy transferred by light (J)	Energy transferred to thermal stores (J)	Wasted energy (%)
Incandescent bulb 60 W	60	6	54	90
Compact fluorescent bulb 15 W	15	11	4	
LED lamp 8 W	8	1.6	6.4	

6 Calculate the percentage of wasted energy for each type of bulb. The value for an incandescent bulb has already been done.

7 Discuss the benefits or disadvantages to society of using the different types of light bulb.

» Extend: Consequences of energy dissipation

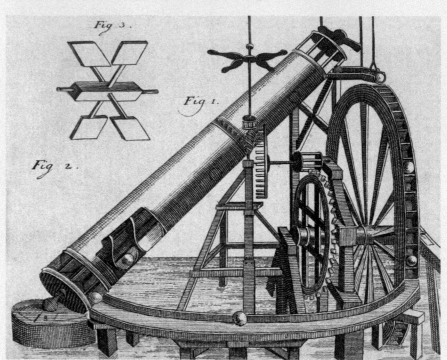

▲ A 1664 design for a perpetual motion machine by Ulrich von Kranach. Cannonballs fall into the large wheel, which turns and drops them onto a slope. They are then lifted up the tube which is linked to the big wheel. To work, the design requires energy to be created from nowhere

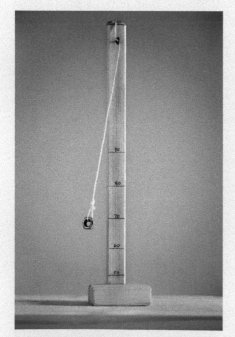

▲ A simple pendulum swings from side to side

There have been lots of people who have claimed to have invented a perpetual motion machine. This is a machine that, once it is set going, will continue to move without any additional energy transfer into the system.

We can understand why these machines never work, if we think about the energy stores for a simple pendulum.

When the pendulum is at its highest point, its store of gravitational energy will be filled. The pendulum then swings towards the centre of its motion, and speeds up. Energy is transferred to a kinetic store. Once the pendulum has passed the middle of its swing, it starts to slow down and get higher. Energy is transferred from a kinetic to a gravitational store.

If there were no energy dissipation, this transfer would keep on going. However, there is friction at the point where the string is attached to the post. The friction dissipates energy by heating. On each swing, there is slightly less energy in each store. The pendulum goes less high, and travels more slowly. Eventually it will come to a stop, hanging straight down.

▲ A bouncing tennis ball showing the position of the ball as it bounces twice

When you drop a ball, it will often bounce back to a lower height than it was dropped from. The photo shows the height of a tennis ball as is dropped and bounces back up.

Tasks

❶ Suggest what energy stores are filled and emptied for a bouncing ball.

❷ Explain why the ball doesn't return to its original bounce height.

❸ A company sells 'super bounce balls'. When dropped, these balls do bounce back almost as high as they are dropped. Suggest why.

Renewable energy resources and dissipation

The Sun is an energy resource. Energy is transferred by light (and other radiation) to the Earth. How much energy is transferred to the surface depends on cloud cover and where you are on the surface.

The amount of energy from the Sun is more than enough to meet the needs of the human population. Solar panels are therefore important as a renewable energy technology.

▲ Solar panels on a roof-top help to reduce the electricity bill for the people living in the house

Solar photovoltaic (PV) panels generate an electric current using the incoming sunlight. This current can be used to charge a rechargeable battery – in other words, a chemical store of energy is filled. The bigger the solar PV panel, the more energy can be transferred.

The percentage of useful energy transferred for a solar PV panel can be as low as 14%. That means that 76% of the energy is wasted, and cannot be used to generate a current.

We can calculate the percentage of wasted energy for other energy resources. This is shown in the chart.

Tasks

❹ How much energy is wasted by hydroelectric power stations?

❺ Suggest why we don't use hydroelectric very much as a way to generate electricity.

❻ Compare the percentage of energy wasted by coal power stations with the percentages for the renewable energy sources in the diagram.

❼ If a solar PV panel receives 300 J of energy, how much useful energy would the panel transfer?

Waves

Learning objectives

7 Sound

In this chapter you will learn…

Knowledge

- that sound is vibrations that travel as a longitudinal wave through substances
- that sound travels faster in denser materials
- that louder sounds have a waveform with a greater amplitude
- that higher-pitched sounds have a greater frequency (and therefore a shorter wavelength)
- that sound cannot travel through a vacuum
- that the speed of sound in air is 330 m/s, and that it is much slower than light
- the definitions of the terms vibration, longitudinal wave, volume (loudness), pitch, amplitude, wavelength, frequency, vacuum, oscilloscope, absorption, auditory range and echo

Application

- how to explain what happens when sound is reflected, transmitted or absorbed by different materials
- how to use the idea of a longitudinal wave to explain how sound travels
- how to describe the amplitude and frequency of a wave from a diagram or oscilloscope picture
- how to use drawings of waves to describe how sound waves change with volume or pitch

Extension

- how to suggest the effects of damaging particular parts of the ear on a person's hearing
- how to evaluate the data behind a claim for a sound creation or blocking device, using the properties of sound waves
- how to use diagrams to compare the waveforms of sounds with different pitches or volumes, for example when a musical instrument is played

8 Light

In this chapter you will learn…

Knowledge

- that some light is absorbed and some is reflected when a light ray meets a different medium
- that for a plane (flat) mirror, the angle of incidence equals the angle of reflection
- to use the ray model to describe the formation of an image in a mirror
- that objects appear different colours depending on the frequencies they reflect
- that when light enters a denser medium it bends towards the normal; when it enters a less dense medium it bends away from the normal
- about refraction through lenses and prisms and how it is shown using ray diagrams
- that light travels at 300 million metres per second in a vacuum
- that different colours of light have different frequencies
- the definitions of the terms incident ray, reflected ray, normal line, angle of reflection, angle of incidence, refraction, absorption, scattering, transparent, translucent, opaque, convex lens, concave lens and retina

Application

- how to explain the colour seen when coloured lights are mixed, or when objects are viewed in different lights
- how to describe how light passes through lenses and transparent materials
- how to describe how lenses may be used to correct vision
- how to draw ray diagrams to show how light reflects off mirrors, forms images and refracts

Extension

- how to predict whether light will reflect, refract or scatter when it hits the surface of a given material
- how to use ray diagrams to explain how a device with multiple mirrors works

7 Sound

» Transition: Surrounded by sound

Sounds are everywhere; music, speech, noisy vehicles and notifications on a mobile phone. Both **pitch** (high or low) and **volume** (loud or quiet) are descriptions that help. But what does the word 'sound' really mean? What is happening?

Imagine a student hitting a cymbal. The cymbal surface moves forwards and back; it shakes or **vibrates**. This makes nearby particles in the air vibrate too. Those vibrations make more particles vibrate and so on. A nearby student hears the sound when parts of their ear are made to vibrate by the air.

Seeing vibration

The cymbal is the *source* of the sound. The part of the ear which moves first, the ear drum, is the *detector*. The material in between is called the *medium*; in this case it is air. Breaking down the process into these three steps can help explain what is happening.

The vibration of the source and detector are easy to check. To see what the medium is doing, it is easier not to use air (an invisible gas).

If one end of a stretched spring is pushed suddenly the movement of each part of the spring, forward then back, is clear. Parts of the spring are squashed then stretched. Each part ends up back where it started; there is no permanent movement of the medium, only **displacement**. The medium has transferred or shifted energy between two vibrating objects. This is called a **longitudinal wave**.

Solids, liquids and gases can all be made to vibrate. The closer together the particles are, the faster and more efficient the transfer of energy between source and detector. But what if there are no particles at all?

▲ These instruments are designed to make particular sounds

Key words

A whistle or squeak is a sound with a high **pitch**. Thunder has a low pitch.

The **volume** of a sound, measured in decibels (dB) measures how loud it is.

A **vibration** is a back and forth motion that repeats in a pattern.

The temporary movement of a medium as the wave travels is called **displacement**.

The displacement of a **longitudinal wave** is along or in line with the direction of wave travel.

★ **See Chapter 6 for more about energy transfers.**

Common error

In most cases, sound doesn't involve the air particles moving along the distance from the source to the detector. Actually, the particles of the medium (air) are displaced for a short time. This is called a wave.

Key word

A **vacuum** is a space with no particles in it, so sound waves cannot travel through it.

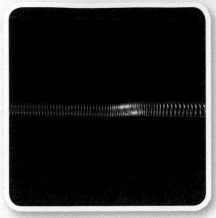

▲ A longitudinal wave travels along the slinky

▲ Three media: air in the jar, glass, then air outside the jar

When the bell rings it makes the air in the jar vibrate. This makes the glass itself vibrate. This makes the air outside the jar vibrate. If the pump is used to remove the air from inside, the **vacuum** does not have enough particles to pass on the vibration properly. The bell seems quieter.

Know >

1 Does a squeaking mouse make a high- or low-pitched sound?

2 Give two examples of a high-volume sound you have heard today.

3 What part of the instrument is the source of the sound made by a guitar?

Apply >>

4 What does the 'bell in the bell jar' experiment tell you about how light travels?

5 A student notices that sounds under water seem louder than he would expect. Explain why, using ideas about particles.

6 Emily and Oscar are arguing about models of sound. Explain what is wrong with each model.

I think sound is like a paper aeroplane flying across a room.

I think it's better to imagine a row of dominoes being knocked over.

Key fact

→ Sound travels at about 330 m/s in air, faster in liquids and faster still in solids.

» Core: Describing sound

Microphones and loudspeakers are familiar devices – mobile phones have at least one of each. In everyday language they convert sound into electricity and back. To a scientist, a microphone is a detector of sound waves in the air. Part of the microphone vibrates, and this movement causes a changing electrical current. This current can be displayed on a screen. In 1897 this allowed scientists to 'see' sound for the first time. A modern **oscilloscope** works in a similar way, and software can be installed on mobile devices to use their microphones for the same task.

> **Key word**
>
> An **oscilloscope** displays electrical signals on a screen. Often these electrical signals were converted from sound waves being detected by a microphone.

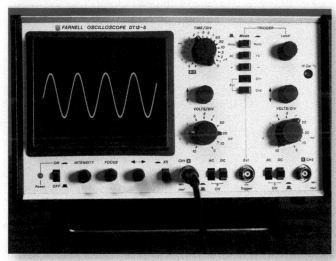

▲ The way oscilloscopes work has changed, but they still do the same job

> **Key words**
>
> The height of the wave, measured from the middle, is called the **amplitude**.
>
> The time taken for one complete wave to pass a point is called the **period**.
>
> The number of complete waves detected in one second is called the **frequency**.

A simple sound, like a whistle, is easiest to interpret. If many sounds are detected by the microphone the picture can look very confusing. There are particular measurements scientists find useful to describe the wave on the screen and the original sound: **amplitude**, **period** and **frequency**.

Amplitude depends on how far the detector has been moved (displaced) by the vibrating air. Louder sounds have a higher amplitude.

The time scale on the diagram on the left is in milliseconds (ms), so the period of this wave is 3 ms. After this time the pattern will be repeated.

If the period is high, the frequency will be low. High-pitched sounds such as beeping and squeaking have a high frequency and a short period. Frequency is measured in Hertz (Hz); a sound at 4000 Hz causes 4000 vibrations each second.

Oscilloscope traces can look complicated. It is best to think of them as a graph showing how the detector is pushed one way, then the other, as the wave goes through it over time. As the wave is absorbed, the detector gains some of the energy that was shifted from the vibrating source.

▲ The line on the oscilloscope screen is sometimes called a 'trace'

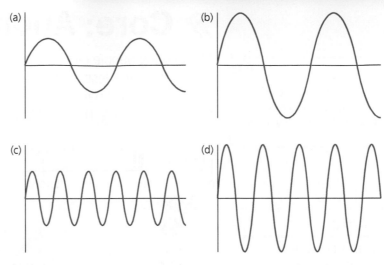

▲ (a) Low amplitude, low frequency. (b) High amplitude, low frequency.
(c) Low amplitude, high frequency. (d) High amplitude, high frequency

Common error

An oscilloscope shows the period of the sound wave, not the **wavelength**. This is a different quantity, and hard to measure for sound waves.

Know >

1 What does the height of an oscilloscope trace measure?

2 What is the unit of a wave's:

a) frequency?

b) period?

3 A loudspeaker vibrates 120 times in 2 s. What is the frequency?

Apply >>

4 In the diagram above, a) represents a quiet, low pitched sound. What do b), c) and d) represent?

5 Give two examples of a high-frequency sound.

Extend >>>

6 Using graph paper, draw an oscilloscope trace for a sound wave with an amplitude of 0.2 A and a period of 20 ms.

7 What is the frequency of this wave?

Key fact

→ The more waves there are on the oscilloscope screen, the higher the frequency.

» Core: Auditory ranges

Most young people can hear sounds between 20 Hz and 20 kHz. If the frequency is less than 20 vibrations per second or more than 20,000 vibrations per second, no sound is detected even though the air is still vibrating. This is called a hearing or **auditory range**.

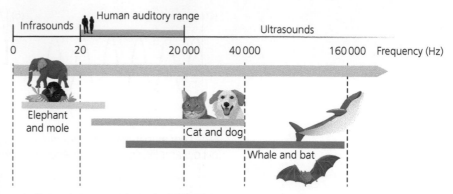

▲ Different animals can hear sounds of different frequencies

Vibrations above 20 kHz are called **ultrasound**. Humans cannot hear them, but some animals can. They are used for medical diagnosis as an alternative to X-rays. Some animals use them for communication and navigation, including whales, dolphins and some species of bat. This usually works by the animal making a series of very high-pitched sounds. All of the **echoes** – reflections of the sound from nearby objects – let them detect what is nearby, how it is moving and even what it might be.

The auditory ranges shown in the diagram are averages. In particular, as humans age they tend to lose the ability to hear sounds at higher frequencies. This is made worse if the ears have been damaged by high-amplitude sounds, perhaps from heavy machinery or very loud music.

» Core: The ear

The sound wave in the air causes the ear drum, a thin layer of skin, to vibrate. This is transmitted to the smallest bones in the human body, which are called the hammer, anvil and stirrup because of their shapes. Then it is the turn of the liquid in the cochlea to vibrate.

The final step that involves vibrations is the movement of tiny hairs on the inside surface of the cochlea. The movement of these hairs is detected by nerve cells and the information then travels to the brain as an electrical signal. Effectively the hairs are acting like the microphone attached to an oscilloscope! Sudden or gradual damage to the hairs means they will no longer respond properly to sounds; this is one of the causes of hearing loss.

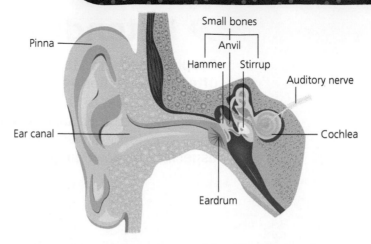

▲ The sound travels through gas, solid and liquid in less than a centimetre

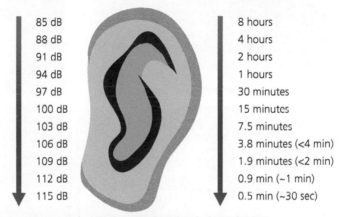

Continuous dB	Permissible exposure time
85 dB	8 hours
88 dB	4 hours
91 dB	2 hours
94 dB	1 hours
97 dB	30 minutes
100 dB	15 minutes
103 dB	7.5 minutes
106 dB	3.8 minutes (<4 min)
109 dB	1.9 minutes (<2 min)
112 dB	0.9 min (~1 min)
115 dB	0.5 min (~30 sec)

▲ Louder sounds will damage hearing in less time

Common error

Students often assume all hearing loss is caused by the ear drum being damaged. This is possible, but rarer than accumulated damage to cochlea.

Know >

1 Starting with the ear drum and finishing with the auditory nerves, list the parts of the ear needed for hearing in order.

2 What is the average upper limit for the human auditory range?

3 Give an example of an animal which can hear lower-frequency sounds than humans.

Apply >>

4 What is the safe time limit for exposure to a sound at 100 dB?

5 A shop puts a speaker outside which broadcasts a sound at 18 kHz. Why do most adults not know what teenagers are complaining about?

Extend >>>

6 Research the use of ultrasound in medicine. Explain why it is better than X-rays for examining the damage to a footballer's ligaments.

» Extend: Insulation

People are so used to the sounds around them that they often end up ignoring them. Students might not be aware of the humming of the classroom heater or the faint buzz of the overhead lights, but that doesn't mean these sounds have stopped. Louder noises are often harder to ignore, especially if they change unpredictably.

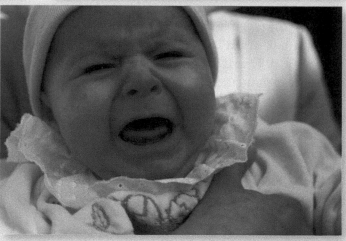

▲ Sources of **noise pollution**

Key words

Sound that is unwanted or annoying is sometimes called **noise pollution**.

During **absorption**, energy is transferred to a material.

The easiest way to reduce the loudness of a sound is to move away. Even a rocket launch is quiet from a large enough distance. If this is not practical, sound insulation is designed to reduce the amplitude by **absorbing** instead of transmitting the sound waves.

The materials chosen are soft, with few flat surfaces. They tend to have lots of gas bubbles because each time the wave is transmitted between solid and gas (or gas and solid) the amplitude is reduced. The materials can be used on walls or ceilings, for example in a house near a wind turbine, or in ear protectors that are worn by a single person. Double-glazed windows also reduce the transmission of sound into a building, even though they may be thought of as mainly thermal insulators.

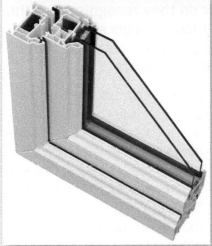

▲ There are many uses of sound insulation materials

Sound can reflect from walls inside a room, especially if there are many hard surfaces. Curtains and carpets reduce this because some of the sound waves are absorbed instead of being reflected. People (and their clothes) also absorb some sound. Sounds echoing in an empty school hall are much louder than those when it is filled with students during assembly.

Engineers investigating sound might use a room like the one in the picture below. It is an anechoic chamber, designed so that very little sound can get in. Almost all of the sound produced inside is also absorbed. They are used to analyse musical instruments and to improve the design of new machines or vehicles.

▲ This room is one of the quietest places on Earth

Key fact

→ The volume of a sound is measured in decibels (dB). The scale is not linear; for each 10 dB increase the loudness doubles.

Tasks

1. Give two sources of noise pollution you have experienced recently.
2. Which will reflect more sound, hard or soft surfaces, and why?
3. Triple glazing has three layers of glass instead of two. Would this work better than double glazing to absorb sound or worse?
4. Give examples of buildings that might install triple glazing.
5. Why might engineers want to reduce the noise produced by a new model of car?
6. A soundmeter is a device which measures volume using a microphone. Describe an experiment to test a range of materials as possible sound insulation.

8 Light

» Transition: Reflection

When toddlers are shown themselves in a mirror they don't always understand. Some will try to look behind the surface. Some make faces and try to play with the child they see. Only as they get older do they recognise that the face in the mirror is theirs. Very few animals are able to pass this 'mirror test', although scientists disagree about how useful it is as a measure of intelligence.

▲ By age two, around two thirds of children recognise their own reflection

How light works is something scientists have been arguing about for centuries. One of the models most useful today is to think of light as a kind of wave. This means some of the ideas in Chapter 7 will be helpful. Other models are useful at other times.

Angles

Just as sound echoes, light reflects from some surfaces. A good starting point is a single source of light in a dark room, pointed at a mirror. If the light is in a narrow enough beam, it is easy to measure the angles of the light before and after the surface. These are called the **incident** and **reflected** rays.

For a mirror, the **angle of incidence** equals the **angle of reflection**. As the below diagram shows, the angles are measured *from* the **normal**. With flat mirrors, this might not seem to matter. If the mirrors are curved then it is more important.

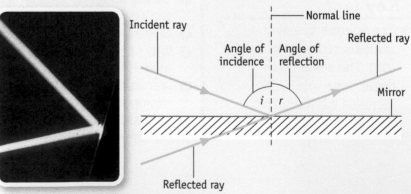

▲ The light is reflected from the mirror surface. The reflected ray seems to come from *inside* the mirror, showing an **image**.

Your knowledge objectives:

In this chapter you will learn:
- that some light is absorbed and some is reflected when a light ray meets a different medium
- that for a plane (flat) mirror, the angle of incidence equals the angle of reflection
- to use the ray model to describe the formation of an image in a mirror
- that objects appear different colours depending on the frequencies they reflect
- that when light enters a denser medium it bends towards the normal; when it enters a less dense medium it bends away from the normal
- about refraction through lenses and prisms and how it is shown using ray diagrams
- that light travels at 300 million metres per second in a vacuum
- that different colours of light have different frequencies
- the definitions of the terms incident ray, reflected ray, normal line, angle of reflection, angle of incidence, refraction, absorption, scattering, transparent, translucent, opaque, convex lens, concave lens and retina

See page 73 for the full learning objectives.

Key words

The incoming ray is called the **incident** ray.

The outgoing ray is the **reflected** ray.

The **normal** is an imaginary line at right angles to the surface. Scientists measure light angles *from* the normal.

The **angle of incidence** is between the normal and the incident ray.

82

Key word

The **angle of reflection** is between the normal and the reflected ray.

The light travels in a straight line until it reaches the mirror. It spreads out a little – this is why a ray box has a narrow slit, so it is narrow at the start – but the ray doesn't bend. The words 'beam' and 'ray' are both used to describe the light and the path it takes. Diagrams are drawn using a ruler to show one possible path for the light, with an arrow to show direction. Often they are drawn to show the edges of a beam or object.

Key fact

→ Sometimes the reflected ray is drawn with two arrows to show that something has happened to it.

Common errors

Students often forget that angles must always be measured from the normal, not the surface.
Rays are drawn with the arrows pointing towards the eye, not away from it. People don't shoot light out of their eyes.

Worked example

Where should Caroline place a mirror so that she could see the coin under the cupboard?

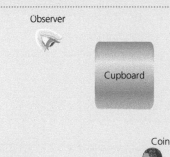

▲ Caroline cannot see the coin through the cupboard

The mirror must be placed so that the incident angle (from the coin to the mirror) and the reflected angle (from the mirror to the eye) are the same.

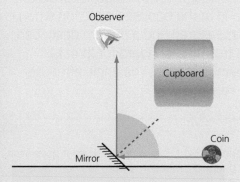

▲ The mirror symbol has been added and the rays of light drawn

Know >

1 A beam of light reflects from a shiny surface. Where do scientists measure the angle of the beam from?

2 What is the light going towards the shiny surface called?

Apply >>

3 Draw the symbol for a mirror and label the shiny side.

4 Look at the diagrams below. Choose the correct letter for the position for the mirror so light is reflected from the object (the black dot) to the eye.

(a)

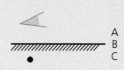
A
B
C

(b) X Y Z

(c) P
Q
R

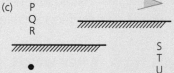
S
T
U

▲ Choose the letter(s) where mirrors should be placed

≫ Core: Transmitted light

Light does not always reflect from a surface. Sometimes it is **absorbed** and sometimes it is **transmitted**. These describe the extremes; there are materials which absorb some light and transmit the rest.

Depending on how a material or surface behaves, scientists use different words to describe it. Clear glass is often considered to be a **transparent** material. Light goes through it without being absorbed. But that doesn't mean that nothing happens to the light. By setting up a ray box so the light enters a glass block, the effect of crossing from one medium to another can be seen.

If the light enters the glass at a right angle (90°), there is no change. The light is travelling along the normal, so the angle of incidence is zero.

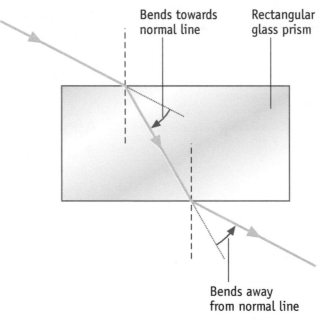

Bends towards normal line

Rectangular glass prism

Bends away from normal line

▲ If the light is not travelling along the normal the direction changes

The diagram above shows that the light bends twice; it bends one way when it enters the glass, and in the opposite direction when it comes out. The angles can be measured and a pattern described. The effect is called **refraction** from a Latin word for 'break', because it can look like the beam is broken. The angle after the surface is called the **angle of refraction**.

▲ The photo has not been edited to look like this; water causes refraction too

Rules for refraction

Refraction happens because the density of the materials is different. Although working out the exact angles is complicated, the pattern is simple. When light enters a denser medium, e.g. from air to glass, it bends *towards* the normal. When it enters a less dense medium, e.g. from water to air, it bends *away from* the normal.

The important thing when studying refraction is to concentrate on one surface at a time. With the glass block the ray is refracted twice. If the shape of the block is more complicated the refracted ray may go in directions that seem surprising. Add the normal to the surface wherever the light enters or leaves and remember the rules about bending towards or away from the normal depending on density.

Key fact

→ Light travels in straight lines unless the medium changes or it meets a surface.

Know >

1 What is special about a transparent material? Give an example.

2 Which direction does light bend when it goes into water from air?

Apply >>

3 Which of these examples is a correct diagram for refraction in a glass block?

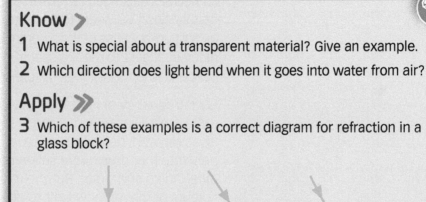

(a)   (b)   (c)

▲ Glass blocks in air; which is correct?

4 Explain the direction of refraction in this triangular prism:

a) going in

b) coming out.

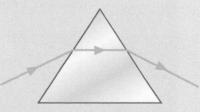

▲ Remember to consider where the normal lines are

Extend >>>

5 Explain why light is said to travel in straight lines although it seems to bend during refraction.

» Core: Colours

Not all materials are transparent. Tinted glass absorbs some of the light instead of transmitting it, which means it is **translucent**. If *all* of the light is absorbed or reflected, and none goes through – for example, a brick wall – it is called **opaque**.

A shiny surface, such as a mirror, reflects the light that meets it in a particular direction. But almost all objects will reflect some light; it's just that it goes in all directions. This is called **scattering** and is part of the reason why people see objects as different colours.

In the same way as sound waves have different frequencies, there are different kinds of light too. Some objects emit (give out) light; they are sometimes described as **luminous**. If it emits a mix of lots of different frequencies it will seem to be white. It turns out that white light can be split up, because of the different parts being refracted at slightly different angles, to make a spectrum. A rainbow is a spectrum which happens when sunlight (a mixture of colours) is separated by raindrops.

Each frequency of light is detected as a slightly different colour; there are millions! One of the first scientists to investigate this, Isaac Newton, believed that seven was an important number so he wrote that there were seven colours: red, orange, yellow, green, blue, indigo and violet. But the reason people still talk about specific colours of the rainbow isn't to do with the physics of light.

The simplest model for seeing light uses three primary colours: red, green and blue. Cells in the eye called cones respond to each of these colours (frequencies) separately. White light is made up of all three colours. Other combinations give the secondary colours: magenta (pink-purple), yellow and cyan (blue-green).

Objects reflect or transmit different mixtures of the three primary colours. A T-shirt that reflects both green and blue light will appear cyan. A filter that transmits red and blue will seem to turn white light into magenta.

▲ How many colours can you see?

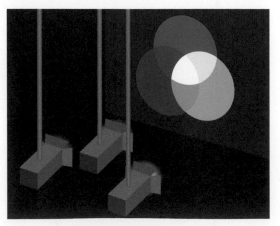

▲ Mixing all colours of light makes white

Worked example

What happens when yellow light is shone through a cyan filter on to white paper?

The easiest way to deal with any questions about mixing and filtering colours of light is to break every colour down into combinations of the three primary colours.

yellow light = red and green light

cyan filter transmits blue and green light
Only the green part of the yellow light is able to go through the cyan filter. This is scattered from the paper, which appears green.

Know >

1 What is the difference between reflection and scattering of light?

2 Give two examples of objects that are:

a) translucent

b) opaque.

Apply >>

3 For each situation, describe the combination of red, green and blue light, with the overall colour if relevant:

a) white light through a magenta filter

b) blue light through a green filter

c) white light scattering from a yellow object

d) magenta light through a red filter and scattering from a cyan object.

Extend >>>

4 Suggest what you would expect if a bright red light was mixed with low-intensity blue light.

» Extend: Sensing light

In Chapter 7 the function of the ear was explained. The part of the body which responds to light is, of course, the eye. This seems very complex, but biologists have shown how it could have developed from a simple patch of light-sensitive cells over generations.

> **Key word**
>
> The **retina** is a layer of light-detecting cells at the back of the eye where an image is formed.

Humans still have that patch of sensitive cells. It's now at the back of the eyeball and is called the **retina**. The different kinds of cells respond to different frequencies or colours of light. In some people, some of these cells are missing or work differently. These people – about one in twelve males and one in 200 females – have what is called colour vision deficiency (CVD) or colour-blindness. Their eyes distinguish between colours in a different way.

▲ The images on the left look the same as the images on the right to people with CVD

▲ Light refracts as it passes through the lens

To focus the light on the retina, other parts of the eye need to bend the light in very precise ways. An important part of this is the lens, which changes shape depending on the muscles that pull on it. The light enters through the pupil – a hole in the coloured tissue called the iris – and is refracted as it goes through the lens.

In the diagram of the eye, the lens is the perfect shape to focus the light on the retina. The two beams from the object edges come together exactly. If the lens doesn't refract them enough, or too much, they will make a fuzzy image. This is why some people need to wear glasses or contact lenses to see clearly.

Key words

A **convex** lens is thicker in the middle and bends parallel light rays towards each other.

A **concave** lens is thinner in the middle and spreads out light rays.

If the person's natural lens doesn't bend the light enough, an extra glass or plastic lens is used which is thicker in the middle. The **convex** shape causes extra refraction towards the middle. Artificial lenses that are thinner in the middle bend light rays away from the middle. These **concave** lenses are good if a person's natural lens is working too hard.

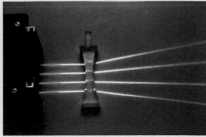

▲ Different shapes of lens have different effects

Tasks

1. How does light get into the eye?
2. Which kind of lens is needed if someone's eye refracts the light too much?
3. A magnifying glass can be used to start a fire by focussing a wide beam of sunlight into a narrow point. What kind of lens is it?
4. Describe what is happening to the beam in the photo below, using the idea of angles to the normal.

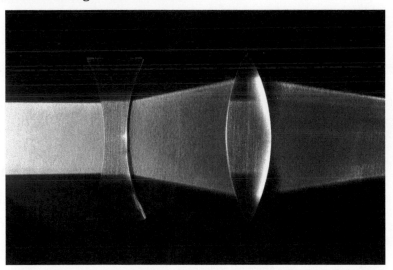

▲ These two lenses do different things

5. Research the common different kinds of colour-blindness. It is possible to find simulations that show the difference that may help. Give examples of difficulties someone with CVD may have during the school day and what could help.

Enquiry:
Heliographs

▲ Pretty for a holiday

Desert island

Nobody wants to get marooned on a desert island (apart from science teachers with really loud classes). But once stuck there, understanding some science can make a difference. Distillation is a way to get safe drinking water from the sea. A magnetised needle will point north for navigation. Shiny surfaces prevent sunburn and heatstroke in the heat of the day. A magnifying glass can focus light to start a fire. But how can a stranded scientist signal for help?

Light from the Sun can be reflected to a passing plane by a shiny surface. This will happen by chance some of the time, so a signalling mirror is moved back and forth to make the light flash in a pattern. This is called a **heliograph**. But where should the mirror be aimed?

❶ Explain why aiming the mirror directly at the Sun, or at the plane, is pointless.

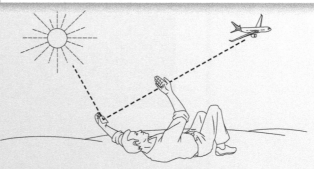

▲ One hand holds the mirror; the other blocks the light to make a flashing signal

Looking for a pattern

While waiting for a plane to arrive, data is collected on how the light is reflected by the mirror.

❷ What device would be needed to measure the angles?
❸ Why repeat the measurements?

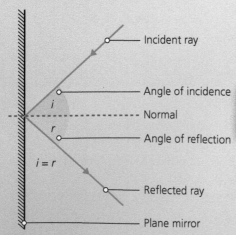
▲ All angles are measured from the normal

Angle of incidence (°)	Angle of reflection 1 (°)	Angle of reflection 2 (°)	Angle of reflection 3 (°)
10	10	11	10
20	21	21	41
30	30	32	31
40	42	41	41
50	53	51	52

4 What is the mean angle of reflection when the angle of incidence is 50°?

5 When the angle of incidence is 20°, the mean result is recorded as 21°. Explain why.

6 What is the relationship between the angle of incidence and the angle of reflection?

Using the pattern

A plane is seen in the sky. The angle between the position of the plane's cockpit and the Sun is 54°.

7 Describe in words where the mirror should be aimed to send a signal.

8 What angle should there be between the pilot in the cockpit and the normal?

The plane passes without noticing the signal. Two more mirrors are found which could be used for signalling. They are not flat, but slightly curved. Could these work better?

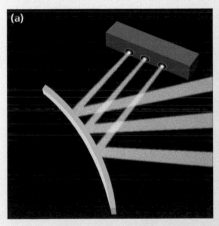 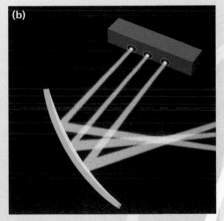

▲ Both mirrors are curved: (a) convex and (b) concave

Like lenses, these mirrors bend parallel rays of light either *towards* or *away from* the normal. This means the convex lens spreads out the light rays while the concave lens focuses them on a point 1 km away.

9 What effect would using each of these as a signalling mirror have? Which of the three choices would be better and why?

Matter

Learning objectives

9 Particle model

In this chapter you will learn…

Knowledge
- about the arrangement and movement of particles in solids, liquids and gases
- that a substance is a solid below its melting point, a liquid above it, and a gas above its boiling point
- the definitions of the terms particle, particle model, diffusion, gas pressure, density, evaporate, boil, condense, melt, freeze and sublime

Application
- how to explain unfamiliar observations about gas pressure in terms of particles
- how to explain the properties of solids, liquids and gases using the particle model
- how to explain changes of state (e.g. melting, condensing) using ideas about particles gaining or losing energy
- how to draw before and after diagrams of particles to explain observations about changes of state, gas pressure and diffusion

Extension
- how to suggest and justify whether a substance which behaves unusually is a solid, liquid or gas
- how to evaluate observations that provide evidence for the existence of particles
- how to make predictions about what will happen during unfamiliar physical processes, in terms of particles and their energy

10 Separating mixtures

In this chapter you will learn…

Knowledge
- that a pure substance consists of only one type of element or compound and has a fixed melting and boiling point
- that mixtures may be separated due to differences in their physical properties
- that air, fruit juice, sea water and milk are mixtures
- the definitions of the terms solvent, solute, dissolve, solution, soluble, insoluble, solubility, pure substance, mixture, filtration, distillation, evaporation and chromatography

Application
- how to use the particle model to explain how substances dissolve
- how to use evidence from chromatography to identify unknown substances in mixtures
- how to choose the most suitable technique to separate out a mixture of substances

Extension
- how to analyse and interpret solubility curves
- how to suggest a combination of methods to separate a complex mixture and justify your choices
- how to evaluate the evidence for identifying an unknown substance using separation techniques

9 Particle model

Your knowledge objectives:

In this chapter you will learn:
- about the arrangement and movement of particles in solids, liquids and gases
- that a substance is a solid below its melting point, a liquid above it, and a gas above its boiling point
- the definitions of the terms particle, particle model, diffusion, gas pressure, density, evaporate, boil, condense, melt, freeze and sublime

See page 93 for the full learning objectives.

» Transition: States of matter

Water is all around us. It can be found on Earth as ice and snow when it is very cold. It can be found as a liquid when it is warmer. And it can be found as a gas when it is very hot. But water is not the only substance that can be found in three states of matter. Almost every pure substance can be turned into solid, liquid and gas if we get the conditions right.

▲ Water is found in three states on Earth, as seen in this iceberg (solid), lake (liquid) and geyser (gas)

Solids, liquids and gases

Solids are things that have a fixed shape. A wooden table is made from wood, which is a solid. Your water bottle is probably made from plastic, which is a solid. The legs of the chair that you are sitting on are probably made from steel, which is also a solid.

When most pure solids are heated, they turn into a liquid. This process is called **melting**. When you are cooking, you may heat some butter in a saucepan. It will melt from a solid into a liquid. A liquid does not have a fixed shape, so it can be poured into a different container. Cooking oil is a liquid, and so is the petrol or diesel that cars use as fuel.

When most pure liquids are heated, they turn into a gas. This process is called **boiling**. When water is heated, it turns into steam. Gases do not have a fixed shape, and they spread out to take up as much space as possible. Air is a mixture of lots of gases, including nitrogen, oxygen and carbon dioxide.

> **Key fact**
> → Everything is made of particles.

Key words

Melting is when a solid turns into a liquid when it is heated.

Boiling is when a liquid turns into a gas when it is heated to its boiling point.

The temperature that a substance melts at is called its **melting point**. The melting point of ice is 0 °C (degrees Celsius). The melting point of chocolate is around 34 °C, which is why it melts in your mouth, which is 37 °C.

The water cycle below describes how water changes state and moves around on Earth. Water **evaporates** from lakes and the oceans when it is warmed by the sun. It cools in the atmosphere and **condenses** to form clouds, which produce rain. In cold conditions, the water falls as snow instead of rain.

▲ Ice melts into liquid water at 0 °C

Key words

Evaporation is a similar but slower process than boiling. It is when a liquid turns into a gas below its boiling point.

Condensing is when a gas turns into a liquid when it is cooled.

Common error

Some people think that water is the only substance that can exist as a solid, liquid and gas! This isn't true. If you look at a candle which is burning, the solid wax melts near the wick to make liquid wax. But the flame is where the wax is so hot that it turns into a gas before burning.

Key facts

➔ Solids have a fixed shape.
➔ Liquids can be poured because they don't have a fixed shape.
➔ Gases spread out to take up as much space as possible inside a container.

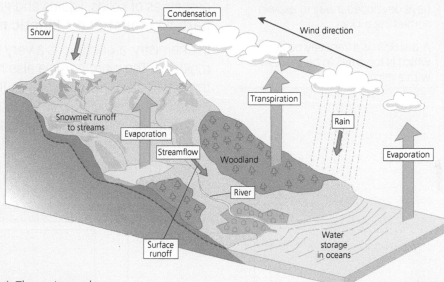

▲ The water cycle

Worked example

Describe what happens to some cooking fat when it is taken out of the fridge and put into a hot frying pan and heated very strongly until it is at a very high temperature.

The fat heats up in the pan and eventually gets to its melting point. This is when it turns into liquid fat. When it gets very hot, it will reach its boiling point, and then it turns into a gas.

Apply »

1 Sort these substances into solids, liquids and gases, when they are at room temperature.

orange squash, chocolate, hydrogen, wood, carbon dioxide, oil

2 Which of these processes explains why a cold window mists up when you breathe on it?

melting, freezing, boiling, condensing

3 Name two changes of state that take place when a substance cools down.

4 Describe how rain forms as part of the water cycle.

95

» Core: A model of solids, liquids and gases

Solids have a fixed shape, but why? We know that liquids, like water, can be poured, but what makes them do this when solid substances do not normally flow? And what explains why gases spread out so easily, to take up all the space in a container?

Scientists have developed a model to help explain the typical properties of solids, liquids and gases. This **model** is called the **particle model**, or the **kinetic model**.

In Chemistry, a **particle** is a very tiny object, normally an atom or molecule. The word particle also has an everyday meaning, but these particles are much bigger, like a grain of sand. When we describe the particles in solids, liquids and gases, we are talking about atoms and molecules. These particles are always moving, which is why the model is sometimes called the kinetic model.

Look at the diagrams below, which show how particles are arranged in a solid, liquid and gas.

	Solid	Liquid	Gas
How are the particles arranged?	• Particles are packed closely together. • Particles have a regular arrangement. • Particles have strong forces of attraction between them.	• Particles are packed closely together with some small spaces. • There is a random arrangement of particles. • Particles have quite strong forces of attraction between them.	• Particles are spread far apart with large spaces between them. • There is a random arrangement of particles. • Particles have very weak forces of attraction between them.
How do the particles move?	• Particles are fixed in place. • They move only by vibrating.	• Particles can move over each other as well as vibrating. • They move randomly.	• Particles move quickly in straight lines. • Particles bounce off each other and the walls of the container. • They move randomly.

The particle model of the three states of matter is useful for us because it helps to explain the typical **properties** of solids, liquids and gases.

Common error

There is nothing in between the particles in a solid, liquid or gas. It is empty space. We sometimes call this a vacuum.

Solids

Solids have a fixed shape because the particles are fixed in place. Solids expand when heated because the extra energy makes the particles vibrate faster so they move slightly further apart. Solids are often good conductors of heat because the vibrating particles at one end of the solid bump into their neighbours, and pass their energy (vibrations) along.

Liquids

Liquids can be poured from one container to another because the particles can move freely over each other. Liquids are incompressible (cannot be squashed) because the particles are mostly touching.

Gases

Gases take up all the space in a container because their particles move very fast until they bounce off the walls of the container. The particles only have very weak forces of attraction between them, and are too light to be affected by gravity.

Worked example

Using ideas about the particle model, explain why solids have a fixed shape and cannot be compressed.

Solids have a fixed shape <u>because</u> they are made from particles which are fixed in place in a <u>regular arrangement</u>. They cannot be compressed because each particle is touching the neighbouring particles, <u>so</u> they cannot be squashed any closer together.

*It is often a good idea to use the word **because** when answering a question which asks you to **explain**.*

Try to use key terms accurately, like this.

*__So__ and **therefore** are both words that can be used to link ideas in an explanation.*

Know >

1 Which state of matter contains particles that are fixed in place?

2 Describe the arrangement and movement of particles in a solid.

3 List the typical properties of a liquid.

Apply >>

4 Describe two differences between the arrangement of particles in a solid and a gas.

Extend >>>

5 Predict whether a lump of solid iron would float or sink in a container of liquid iron.

6 Water is considered to be a very unusual substance because it expands when it freezes. Use the particle model to explain why this is so unexpected.

» Core: Changes of state

A pure substance can change from one state to another. This is usually caused by a change in temperature. These changes have specific names, as shown in the diagram below.

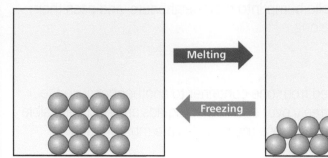
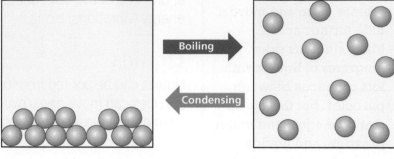

▲ Using the particle model to explain changes of state

Melting and freezing

When a pure substance is heated, the particles gain energy and vibrate more quickly in their fixed positions. Eventually, they gain enough energy to break free from their neighbours and they can then move over each other. The solid turns to a liquid and this process is called melting.

A solid melts at a specific temperature, which is called its **melting point**. Different pure substances have different melting points, as you can see from the table below. Adding another substance will change the melting point, so when salt is added to ice it lowers the melting point, often causing it to melt.

Substance	Melting point (°C)
Ice	0
Copper	1085
Sodium	98
Methane	−183
Mixture of salt and ice	−21 to −1

When you cool a liquid down below its melting point, it will turn into a solid. This process is called **freezing**.

Boiling and condensing

When a liquid is heated, the particles gain energy and move around more quickly. Eventually, they gain enough energy to escape from the liquid and the liquid turns into a gas. This process is called boiling, and it takes place at a temperature known as the **boiling point**. Adding impurities to a liquid changes its boiling point.

When you cool a gas down below its boiling point, it will turn into a liquid. This is called **condensing**. Condensation forms on cold surfaces when they are near to warm gases, such as a cold window in the bathroom when you are taking a shower.

Key words

The **melting point** of a solid is the temperature at which it melts when it is heated.

Freezing is when a liquid turns into a solid as it cools down.

Common error

Lots of people use the word 'freezing' only for water. It is fine to use it for other substances, too. The word 'solidify' is OK to use as well.

Key words

The **boiling point** of a liquid is the temperature at which it boils when it is heated. It is usually the maximum temperature that a liquid can be heated to.

Condensing is when a gas turns into a liquid.

Evaporation

Evaporating is a similar process to boiling, but it takes place more slowly, and the liquid does not have to be heated to its boiling point. Water evaporates from lakes and oceans during the water cycle, but the water is well below 100 °C! Evaporation only occurs at the surface of a liquid, but boiling can occur throughout a liquid, especially if it is heated from below. This is why you can see bubbles rising through a pan of water being heated on a stove to its boiling point.

Key word

Evaporation is when a liquid slowly turns into a gas at a temperature that is below its boiling point. Molecules escape from the surface when they have enough energy.

Worked example

The boiling point of zinc is 907 °C. Its melting point is 420 °C. Work out what state zinc is in at the following temperatures: 100 °C, 500 °C, and 1100 °C.

It is important to recognise that this question has three parts and will require an answer that is in three parts.

When a substance is colder than its melting point, it is a solid, so zinc is a solid at 100 °C. When it is heated to a temperature above its melting point, it turns into a liquid, so zinc is a liquid at 500 °C. When the zinc is heated above its boiling point, for example 1100 °C, it turns into a gas.

Know >

1 What term is given to the temperature at which a substance turns from a solid into a liquid?

2 State the boiling point of water, in degrees Celsius.

3 State what happens to liquid mercury when it is supplied with thermal energy.

4 Describe how the arrangement and movement of the particles changes when steam condenses into liquid water.

Apply >>

5 Explain why salt is added to the roads in the UK in winter.

6 Aluminium has a melting point of 660 °C and a boiling point of 2470 °C. Work out what state aluminium is in at the following temperatures: 1200 °C, 100 °C, 3100 °C, 0 °C, and −200 °C.

Extend >>>

7 Suggest why adding salt to the roads would not work in Siberia (northern Russia) in the winter.

8 Suggest an approximate boiling point for oxygen. Explain your answer.

9 Xander suggests that hot water will freeze more quickly than cold water. His friends do not believe him. You are going to help to design an investigation that will enable Xander to see if his suggestion was correct.

a) Suggest why Xander's friends do not believe him, using ideas about energy and changes of state in your answer.

b) Describe a simple method that Xander could use in his investigation, using a freezer and some ice cube trays.

c) List some variables that Xander would have to control in his investigation in order to make it a fair test.

» Core: Sublimation, diffusion and pressure

Pressure

Pressure is the effect of a trapped gas or liquid which is pushing out on the walls of its container. You can imagine that when you blow more air into a balloon, there is more gas pushing on the inside of the rubber, so the **gas pressure** is increasing. The same thing happens when you pump up a tyre.

Gas pressure occurs because the gas particles are moving very quickly and colliding with the walls of the container. In a balloon, the more particles of gas there are trapped inside, the more frequent the collisions, so the higher the pressure. If the balloon is heated, the gas particles trapped inside move much faster and collide more often with the sides of the balloon, so that increases the pressure as well.

The image shows a demonstration of pressure. A peeled boiled egg rests on top of a conical flask that contains some gently boiling water. As the steam is heated, the pressure increases and this lifts the egg off a little, so some steam escapes. When the steam in the flask cools, its pressure decreases and the neck of the flask is sealed by the egg. But then the **air pressure** above the egg is greater than inside the flask, so the egg is pushed into the neck of the flask by the normal air pressure above.

Pressure can also affect boiling and condensing. A gas can be turned back into a liquid at a temperatures above its boiling point by compressing the gas. This will force the gas particles to move closer together until most of them are touching, and then the gas will have become a liquid.

If the pressure of the air around a liquid decreases, it makes the liquid more likely to boil, even if it is below its boiling point. The air pressure on Mount Everest is much lower than at your school, so water boils at a lower temperature on top of Everest; as low as 90 °C.

Sublimation

Some solids do not melt when they are heated at certain pressures. They turn straight from a solid to a gas. This is called sublimation. Solid carbon dioxide (called dry ice) **sublimes** if it is warmed above −79 °C.

▲ A bicycle tyre being inflated using a pump

Key words

Gas pressure is the force pushing outwards on the walls of a container, which is caused by the gas particles colliding with the container walls.

Air pressure is caused by the weight of the atmosphere above an object or surface.

▲ Using air pressure to push a boiled egg into a conical flask

Key words

Some substances do not commonly exist as a liquid and **sublime** straight from a solid to a gas when heated.

▲ Dry ice is solid carbon dioxide

Diffusion

Have you ever noticed that when someone sprays deodorant a few metres away from you in the changing room, you can't smell it at first but then after a minute or so you can? The deodorant particles are **diffusing** through the air, and spreading out from their original source. Eventually, the smell is evenly spread throughout the changing room.

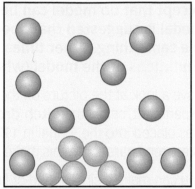 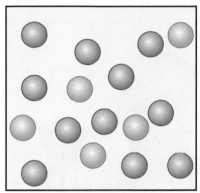

| At the start | After diffusion |

▲ How diffusion occurs in gases

Worked example

On Open Evening, Ms Scutt did her favourite demonstration. She poured some liquid nitrogen into a small plastic bottle and screwed the lid on tight. She then dropped this into a bowl of warm water and ran away a safe distance. Liquid nitrogen has a boiling point of −196 °C. Predict what will happen and explain your answer.

I predict that the small plastic bottle will explode. This is because the warm water will give the nitrogen particles lots of energy. The liquid will boil quickly. The gas particles will push on the walls of the bottle until the pressure is so great that it will cause the bottle to explode.

Answer the prediction part of the question first.

*Remember that explanations can often use the word **because**.*

Use key words.

Know >

1 What happens to the pressure as you climb a mountain?

2 What name is given to the process when one substance moves through a gas or liquid to spread out?

Apply >>

3 Iodine is a grey solid at room temperature. When heated gently with a Bunsen burner it produces a purple gas. Seb's teacher told his class that iodine sublimes when heated. Seb decided to look on Wikipedia and found that the melting point of iodine is 114 °C and the boiling point is 184 °C. Do you think that the iodine was subliming? Explain your answer.

Extend >>>

4 If you put a drop of food colouring onto the surface of a bowl of jelly it spreads through the jelly. But it spreads very very slowly indeed, and may take hours or days to move even a few millimetres. In a bowl of water, the food colouring would move much faster through the water. Why is the movement so slow through the jelly?

» Extend: Limitations of the particle model

When scientists develop a model, like the particle model of solids, liquids and gases, they hope that other people will find it useful to help to explain things. However, scientists accept that no model can be perfect. Sometimes, a new model is suggested that does a better job of explaining the same thing. Other times, they simply acknowledge the limitations of the model (what it is unable to explain).

Have a look at the picture on the left. It is a very famous experiment, called the **pitch drop experiment**. The pitch (or tar) was placed into the funnel in 1927 and it drips approximately once every ten years. Does this make it a liquid?

You can make a similar kind of fluid yourself. Mix cornflour with a little water using a strong wooden spoon. You will find that if you stir it slowly, it acts like a liquid. If you stir it fast, it acts like a solid and cracks into pieces. You can even smash it with a hammer. But afterwards the pieces flow back together like a liquid.

> **Key word**
>
> **Limitations** are the weaknesses of a model.

▲ The world-famous pitch drop experiment

> **Key word**
>
> **Density** is the mass of a substance divided by its volume. Density may have units of kg/m³, or g/cm³.

▲ Nitrogen dioxide gas is very dense

▲ Cornflour and water paste. It acts like a liquid when it is stirred slowly, but acts like a solid when stirred quickly

Some gases have a very high **density**. This means that a specific volume of the gas has a high mass, compared with other gases. One example is nitrogen dioxide (NO_2). Nitrogen dioxide is a toxic and corrosive brown gas. Because it is denser than air, it stays in the bottom of this beaker, and can be poured out of the beaker, just like a liquid. However, if you leave it long enough it will diffuse and fill up all the space available to it, like all gases.

Brownian motion

In 1827, the plant biologist Robert Brown was looking at pollen grains in water through a microscope and he noticed the tiny pollen particles moving in the water in a jittery motion. He couldn't explain why they were moving. It was not until 1905 that Albert Einstein proposed an explanation for the random movement. These pollen particles were being jostled about by the random collisions of even smaller particles (molecules of water) which were too small to be seen, even using a very powerful microscope.

Imagine a huge crowd of people at a concert, jumping up and down. Now imagine dropping a huge beach ball into the crowd. Imagine you are watching from high above the crowd in a helicopter. Can you predict what would happen to the beach ball? Would it stay still? In this analogy (or model) the beach ball is the small particle that Brown was watching, and the people are the molecules of water, too small to be seen.

▲ Robert Brown, 1773–1858

Tasks

Why is it called 'Brownian motion'?

Robert Brown was not the first person to observe tiny particles moving randomly under a microscope. In 1785, forty-two years earlier than when Brown observed his pollen grains, a Dutch scientist called Jan Ingenhousz observed a similar thing. Ingenhousz was looking through a microscope at particles of coal dust on the surface of alcohol. He noticed that they moved around in an unpredictable and irregular way.

1. Use ideas about Brownian motion to explain why the coal dust particles were moving on the surface of the alcohol.
2. Suggest why Ingenhousz was not given the credit for how important his coal dust observations would later turn out to be. Why was it called 'Brownian motion', and not 'Ingenhousz motion'?
3. Why do you think that it wasn't called 'Einsteinian motion'?

▲ Jan Ingenhousz, 1730–1799

Slime

You might have had a chance to make slime at school or at home. You can do this by mixing white PVA glue with a solution of laundry starch or borax. Slime can be stretched between your hands, rolled into a ball, bounced on the floor, poured from one container into a puddle (very very slowly) and does not evaporate if left for three days.

4. What properties from those listed above, or your own experience of slime, suggest that it is a liquid?
5. What properties suggest that it is a solid?
6. Do you think that slime should be classified as a solid or a liquid? Justify your decision.

Further research

Find out the meaning of the terms 'triple point', 'critical point' and 'supercritical fluid'.

Enquiry:
Solid, liquid or gas?

Key words

A **secondary source** is a source of data which was not collected by your own experiment. This could include a data book, or someone else's research.

A **continuous** variable has values that can be any number.

A **discontinuous** variable has values that are words or discrete numbers.

A **variable** is something that can be changed, measured or controlled.

The **independent variable** is the factor you change in an investigation.

The **dependent variable** is the factor you measure in an investigation.

Graphs can help us to spot patterns and identify trends in data that we have collected ourselves, or found from **secondary sources**.

Data can be described using lots of key words.

Continuous data is a set of data where the values can be any number. It does not need to be an integer (whole number). It will usually have units and these should be included on the labels for an axis on a graph. Examples of continuous data in science include length, mass, temperature and electrical current.

Discontinuous data is when the values are words or discrete numbers. One example is shoe size, because you can be size 5, or a size 5½, or a size 6. But you cannot be a size 5.2. Another example is hair colour. Sometimes you hear discontinuous data being called **categoric data**, because it describes different categories, e.g. hair colour.

In an investigation, the **variable** that you choose to change is called the **independent variable**. This always goes on the x (horizontal) axis of a bar chart or line graph. The variable that you decide to measure is called the **dependent variable**, because its value usually depends on the change you make to the independent variable. The dependent variable goes on the y (vertical axis) of a bar chart or line graph.

Different sets of data usually need to be plotted on different graphs or charts.

Pie chart

This gets its name from the shape of a pie when it is cut into pieces, and it shows the percentages of different categories. The whole circle (or pie!) is 100%. For example, a pie chart can be used to show what percentage of our electricity comes from different energy resources.

Bar or column chart

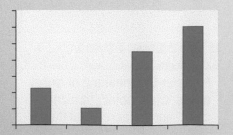

This is used to display the numerical values of different categories. For example, a bar chart can be used to to show what percentage of people in a class have brown, blue or green eyes.

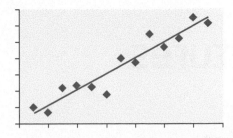

Scatter graph or line graph

This is the graph to use when both the independent variable (*x* axis) and the dependent variable (*y* axis) are continuous. It shows whether there is a **correlation** between the two variables. If there is a clear pattern, and a **line of best fit** can be drawn which shows this trend, then we call the graph a **line graph**. The line of best fit could be straight on some graphs and curved on other graphs, as long as it follows the trend in the data points. Without a line of best fit, it is a scatter graph. For example, a line graph would help you to investigate whether there is a relationship between the force on a spring and its length.

Representing data on melting points and boiling points

Look at the graph below, which shows the melting and boiling points for some different substances. We can call this type of bar chart a 'stacked bar chart'. For each substance, the melting point is shown at the top of the green bar and the boiling point is shown at the top of the red bar. The temperatures in the table and on the graph are measured in units called kelvins, K. This scale of temperature has no negative values, because zero kelvin is the lowest possible temperature, called absolute zero.

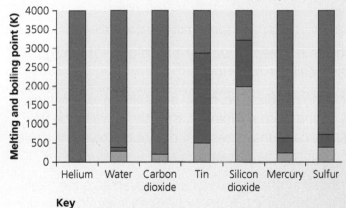

Substance	Melting point (K)	Boiling point (K)
Helium	1	4
Water	273	373
Carbon dioxide	195	195
Tin	505	2875
Silicon dioxide	1986	3220
Mercury	234	630
Sulfur	388	718

▲ A stacked bar chart showing the melting and boiling points of a variety of substances

For each of the following questions, also say whether it is easiest to find the answer from the table or from the graph.

1. Which substance has its melting point very close to 500 K?
2. Which substance has the highest boiling point?
3. Which substance is almost always a gas?
4. Which substance is a liquid over the greatest temperature range?
5. Which substance has the highest melting point?
6. Which substance cannot exist as a liquid according to the data presented here?

Key words

Dissolving is when a substance mixes completely with a liquid called a solvent.

Molecules are tiny particles of a compound. Molecules are made from atoms which are strongly bonded together.

A substance is **soluble** in a particular liquid if it dissolves in it to make a solution.

A **solution** is a mixture made from a liquid and a substance dissolved into it (usually a solid).

A substance is **insoluble** in a particular liquid if it will not dissolve into the liquid.

In **physical changes**, the properties of a substance change but no new substance is made. Changes of state and dissolving are all examples of physical changes.

In **chemical reactions**, a new substance is made because chemical bonds are broken and made.

» Transition: What is a mixture?

Do you like sugar in your hot drinks? When you add sugar to a hot drink and stir it, the sugar dissolves. The crystals of sugar break into tiny particles called molecules. The sugar molecules spread out between the water molecules in the hot drink. All of the hot drink now tastes sweet.

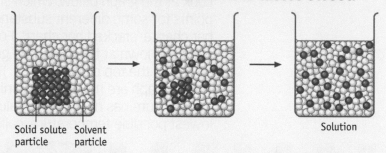

Solid solute particle   Solvent particle   Solution

▲ When a solid dissolves, the particles spread out throughout the liquid

The sugar dissolves in water, so we say it is **soluble**. The mixture of water and sugar is called a **solution**.

Other substances do not dissolve. If you add sand to water, it will never dissolve. If sand were soluble in water, there would be no sandy beaches! We call substances that do not dissolve in a liquid **insoluble**. Sand is insoluble in water.

Dissolving is a **physical change**, so it can be reversed quite easily, just like melting and boiling. **Chemical changes** (chemical reactions) such as burning are usually difficult or impossible to reverse.

▲ Sugar dissolves in hot drinks to make them taste sweet

▲ Beaches are proof that sand is insoluble in water

To get the solid back from a solution, we need to heat up the solution to evaporate the solvent. Filtering the solution will not work because the molecules of sugar are so small that they pass through the filter paper.

If you heat a solution of salt over a Bunsen burner, all the water will eventually evaporate and you will be left with salt crystals.

▲ Evaporating the water from a solution of salt will leave the salt behind as crystals

Worked example

Copper sulfate is a blue crystalline solid. When it is added to water in a beaker, the crystals sink to the bottom. After stirring for one minute, the crystals are no longer there and the whole liquid is pale blue. Explain what has happened.

The copper sulfate has dissolved in the water. Copper sulfate is soluble in water. A solution of copper sulfate has been made. The crystals have broken up into tiny particles and spread out between the water molecules.

Apply >>

1 Cobalt chloride is a pink solid. It is soluble in water. Describe what you would expect to see when crystals of cobalt chloride are added to water and stirred.

2 Describe how you could get solid potassium chloride back from a solution of potassium chloride.

3 What word can be used to describe substances that do not dissolve in a particular liquid?

4 Explain why filtering a solution of copper sulfate will not allow you to separate the copper sulfate from the water.

» Core: Pure substances and mixtures

Pure and impure

A **pure** substance contains only one **compound** or one **element** (see Pupil's Book 2, Chapter 10). If another element or compound is present in small quantities, that extra substance is called an **impurity**.

Common error

In everyday language, people use the word **pure** in a different way. For example, drinking water may be described as **pure**, but it actually contains many dissolved substances.

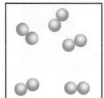   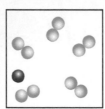  

| Pure element | Pure compound | Pure element | Impure element | Impure compound | Impure element |

▲ Diagrams of pure and impure elements and compounds

A **mixture** is formed when there are two or more elements or compounds present in a sample. The properties of a mixture are different from the properties of the individual substances. Mixtures do not have a chemical formula because the elements and compounds that make up a mixture are not in fixed proportions. Air is a mixture of different gases. Fruit juice and sea water are mixtures of lots of compounds dissolved in water. Milk is a mixture made from water, fat droplets and lots of other chemicals dissolved in the water.

It is usually easier to separate the chemicals in a mixture than it is to separate the chemicals that are bonded together in a compound. Separating mixtures usually requires physical processes, such as boiling or condensing. Separating compounds into their elements requires chemical reactions, like **electrolysis**.

▲ Air, sea water, milk and fruit juice are all mixtures

Key words

A **solvent** is a substance (normally a liquid) which can dissolve another substance.

A **solute** is a substance (normally a solid) which can dissolve into a solvent.

Solubility is the maximum mass of a solute that will dissolve in a specific volume of solvent.

» Core: Solutions

As we have already seen, a solution is a type of mixture. In a solution, one substance (usually a solid, but sometimes a gas) dissolves in a liquid. The liquid is often water. The liquid is called the **solvent**. The substance that has dissolved is called the **solute**. Sea water contains sodium chloride (common salt) dissolved in water. The solvent is water and the solute is sodium chloride.

Solubility

A substance may dissolve in one solvent but not in another. If something dissolves in a solvent, we say that it is soluble. If it doesn't dissolve, then it is insoluble. Sodium chloride is soluble in water, but it is insoluble in another solvent called propanone. Wax is insoluble in water, but it is soluble in propanone. The **solubility** of a solute is how much of it will dissolve in a specific volume of solvent. If the solubility is very low or zero, we describe that substance as insoluble.

Worked example

White potassium chloride crystals are soluble in water. White calcium carbonate powder is insoluble in water. Describe and explain what would be seen when a spatula of each of these solids is added to a separate sample of water.

When potassium chloride is added to water and stirred or shaken, the crystals seem to disappear and a solution is formed. This is because the crystals break into tiny particles and spread out evenly through the water. When calcium carbonate is added to water and shaken or stirred, the water turns white and the solid does not dissolve. This is because it is insoluble. If the mixture is left to stand, the solid powder will settle out at the bottom.

Know >

1 Which of these substances are elements and which are compounds: water, iron, sulfur, carbon dioxide, sodium chloride, hydrogen?

2 Which of these are definitely mixtures and which could be pure substances: air, magnesium, oxygen, milk, sea water, sulfur dioxide, fruit juice?

3 What name is given to a mixture of a solute and a solvent?

Apply »

4 Look at the diagram. Decide whether it shows an element, compound or mixture, whether it is pure or impure, and whether it is a solid, liquid or gas.

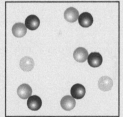

5 A student is asked to find out for homework what steel is made from. He searches online for a chemical formula for steel but cannot find one. Suggest why.

Extend »»

6 Sand is insoluble in water. Salt is soluble in water. A chemist wants to remove the sand from a sample of salt. Describe what she should do to get a pure sample of dry salt crystals.

Enquiry »»»

7 Two samples of drinking water are both described as being the purest water available to buy. Describe how you could find out which one contains the most solutes/impurities.

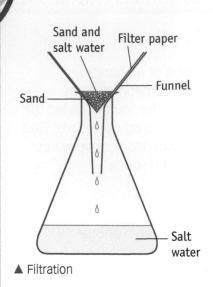

Sand and salt water

Filter paper

Funnel

Sand

Salt water

▲ Filtration

Key words

Filtration is the way to separate an insoluble solid from a liquid.

The **filtrate** is the liquid that passes through a filter.

The **residue** is the solid that is collected in the filter paper.

» Core: Purifying liquids

Filtration

Some liquid mixtures contain particles of an **insoluble** solid. The insoluble solid can be separated from the liquid by **filtration**. The mixture is passed through a piece of filter paper which is supported inside a filter funnel. The liquid passes through the filter paper and is called the **filtrate**. The insoluble solid remains in the filter paper and it is called the **residue**.

The filter paper contains very small holes which are too small to see. The particles of insoluble solid are much larger than the holes, so they do not pass through the filter paper, but the tiny molecules of the liquid do pass through.

Common error

Filtration will not separate a solute from the solvent in a solution. Solutions pass through the filter paper easily because the particles of the solute are much smaller than the holes in the paper.

Evaporation

Evaporation is used to obtain the solute in a solution. When a solution is heated up, the solvent evaporates, leaving the solute behind. The solution is put into an evaporating basin and heated gently using a Bunsen burner until most of the solvent has evaporated. Crystals are left to form and then dried in an oven, desiccator or on the radiator.

Step 1

Step 2

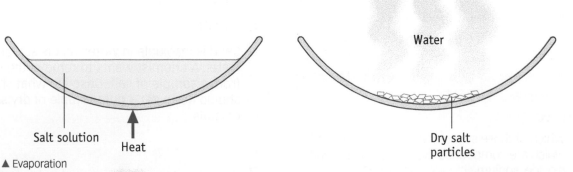

Water

Salt solution

Heat

▲ Evaporation

Dry salt particles

Key word

Distillation is a method of separating a mixture by boiling and then condensing the gas.

Simple distillation

Distillation is a technique for purifying the solvent from a solution. When the solution is heated, the solvent boils off as a gas. This gas is cooled down using a condenser and turned back into a liquid.

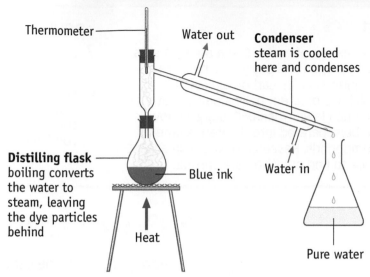

Thermometer

Water out

Condenser
steam is cooled
here and condenses

Distilling flask
boiling converts
the water to
steam, leaving
the dye particles
behind

Blue ink

Water in

Heat

Pure water

▲ Distillation

Fractional distillation

Crude oil is a mixture of liquids that have different boiling points. It can be separated by **fractional distillation**. The mixture is heated up gently until the boiling point of the first liquid is reached. This liquid boils off and condenses in the condenser as the first **fraction**, whilst the temperature of the mixture is held constant. Once all of the first fraction has been separated, the temperature of the mixture is increased until the second fraction boils off and is condensed to form the second fraction. This process can be repeated at higher temperatures to separate all of the fractions in the mixture.

Key words

Fractional distillation is the method used to separate a mixture of liquids which have different boiling points.

A **fraction** is a liquid obtained from fractional distillation. If it is a pure liquid, it will have a specific boiling point. If it is a mixture of liquids, their boiling points will be similar.

Worked example

Explain why heating a solution of copper sulfate in an evaporating basin over a Bunsen flame will not help you collect the solvent. What process must be used instead?

Heating the solution in an evaporating basin will cause the water to evaporate into the air and it will be lost. It cannot be collected and condensed this way. Distillation should be used instead.

Know >

1 What name is given to the liquid produced after filtration has taken place?

2 What process is used to obtain the solid solute from a solution?

3 What process is used to obtain the solvent from a solution?

Apply »

4 One of the important flavourings in cherry cola is a liquid that has a boiling point which is higher than water. State how the water and cherry flavouring can be separated.

5 A power station removes water from a nearby river to use as a coolant. The water contains some sediment which is made up of insoluble compounds. This sediment must be removed before the water can be used in the power station. State the name of the technique which is used to remove the sediment.

Extend »»

6 Air is a mixture of gases. Some are elements, such as nitrogen and oxygen. Others are compounds, such as carbon dioxide. The gases in air all have different boiling points, but they are all much lower than 0 °C. Suggest how air can be separated into different elements and compounds. Research online or in a GCSE textbook to find out if you are correct.

Enquiry »»»

7 List the apparatus that would be needed to separate a mixture of salt and sand.

8 When connecting a condenser to the cold water supply, it is important to put the cold water coming in at the bottom of the condenser. Why is this?

» Core: Chromatography

Chromatography can be used to separate certain types of mixture that would be hard to separate using the methods described already. The most familiar example is using paper chromatography to separate the different pigments (colours) in a food colouring. A small spot of the food colouring is placed near the bottom of a piece of chromatography paper and allowed to dry. The piece of paper is placed into a beaker and water moves up the paper. As the water rises, the pigments in the food colouring dissolve and they rise up the paper too, but at different speeds, so the pigments spread out.

Key word

Chromatography can be used to separate different coloured substances in a mixture.

▲ Paper chromatography

Procedure

1 Draw a pencil line horizontally on the piece of chromatography paper, 1.5 cm from the bottom. If you use pen for this line, it could separate into different pigments.

2 Put a small spot of the mixture onto the line and allow it to dry. You can add more than one mixture if your paper is wide enough.

3 Gently lower the piece of paper into a beaker which contains water to a depth of 1 cm. If the water is too deep it will wash the coloured spots off the paper.

4 Make sure the paper doesn't touch the sides of the beaker, and make sure the beaker stays still on the table.

5 After the pigments have separated but before any of them reach the top of the paper, gently remove the paper from the beaker and allow it to dry.

The speed that a substance moves up the paper during chromatography depends on how strongly it is attracted to the solvent (water) and how strongly it is attracted to the paper. If it is strongly attracted to the water then it will move quickly. If it is strongly attracted to the paper then it will move slowly.

Identifying substances by chromatography

Chromatography can be used to identify a substance in a mixture by comparing it with a known sample. For example, food inspectors are worried that two brands of brown food colouring both contain a red pigment that has been banned because it is dangerous. Paper chromatography is used to separate the pigments in the brown food colourings, but on the same piece of paper the inspectors put a spot of the banned red pigment.

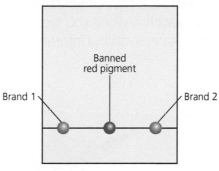
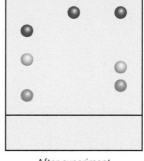

Banned red pigment

Brand 1   Brand 2

Before experiment   After experiment

▲ Before and after the chromatography experiment

The red pigment in the first brown colouring does not match the height of the banned red pigment, so it does not contain the banned pigment. However, the second brown colouring does contain the banned red pigment.

Worked example

Describe how chromatography can be used to identify whether there are any impurities in a sample of green food colouring which is described as 'pure'.

Take a rectangular piece of chromatography paper and draw a pencil line horizontally, 1.5 cm up from the bottom edge. Place a small spot of the green food colouring on the pencil line and allow to dry. Support the piece of paper vertically in a small beaker so that it doesn't touch the sides. Carefully add water using a pipette so that the level of the water is below the line, approximately 1 cm deep. Leave the beaker undisturbed until the water has risen to almost the top of the paper. If the food colour is pure, there will be a single spot of green colour part way up the paper. If there is more than one spot, or a variety of different colours, then the colouring was impure.

Know >

1 What should be used to draw the horizontal line when doing chromatography?

2 Which is the most common solvent used for chromatography in schools?

Apply >>

3 A student thinks that his red pen ink contains only one pigment but his friend disagrees. Describe how they can find out who is correct, including a description of the results that would support each of the students' viewpoints.

Extend >>>

4 Paper chromatography can be done on a circular piece of filter paper by putting the spot of the mixture of pigments in the middle of the filter paper. Water is then slowly dripped onto the spot and spreads out. Suggest why this cannot be used to identify whether a banned pigment is present in the food colouring.

Enquiry >>>>

5 Suggest what might happen if the beaker is moved from one table to another during the chromatography process.

» Extend: Solubility curves

The solubility of a solute is the maximum mass of solute that will dissolve in a specific volume of solvent. Solubility is usually temperature dependent. This means that as the temperature changes, the solubility of the solute changes too. We can do primary or secondary research to find out how the solubility of different solutes changes with temperature. **Primary research** is when you do an investigation yourself and collect data to analyse. **Secondary research** is when you look for other people's experimental data, for example on the internet or in a textbook or databook.

▲ Investigating solubility

Tasks

1 What are the advantages of doing primary research instead of secondary research?
2 What are the advantages of doing secondary research instead of primary research?
3 Lots of information is available on the internet. What advice would you give to a younger student who was looking for **reliable** secondary data online?

Tasks

Primary research

Plan an experiment to investigate how the solubility of sodium sulfate is affected by the temperature of the solvent (water).

4. What is your independent variable and how will you measure it (what with, and in what units)?
5. What is your dependent variable and how will you measure it?
6. List the apparatus you would use.
7. Describe a brief method that will allow you to collect accurate and valid results.
8. What safety precautions should you take?
9. Draw up a results table that you could use.

Secondary research

A student decided to check her results by comparing them with data she found in a data book. The data from a data book is shown below.

Temperature (°C)	Solubility (g per 100 g of water)
10	9
20	19
30	42
40	48
50	46
60	45
70	44
80	43
90	42

10. Suggest why there are no results for 0 °C or for 100 °C.
11. Plot a graph of the solubility against the temperature. You will need to decide which variable goes on which axis and don't forget to label your axes.
12. Why would it be inappropriate to draw a straight line of best fit through the points on this graph?
13. Assume that the results are all accurate and draw a curved line which follows the pattern of the data points on this graph.
14. Describe how the solubility of sodium sulfate is affected by the temperature of the water. Use as much detail as you can.
15. Not all solids follow the same trend as sodium sulfate. Do some secondary research to find out how the solubility of sodium chloride and barium nitrate are affected by temperature.
16. Gases usually become **less** soluble as the temperature increases. We all know that fizzy drinks go flat after they have been left out of the fridge on a warm day. Explain why a fizzy drink from a fridge feels more fizzy in your mouth than one served at room temperature.

Enquiry:
A fraudulent last will

Miss Jackson was an old woman who became very rich because she invented a new way to purify perfume ingredients from plants. Miss Jackson had no children, but had three nieces and one nephew. She loved her nephew the most, because he helped to look after her when she was very old.

How did she extract chemicals from plants?

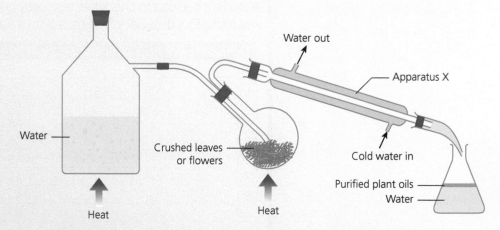

Water out

Apparatus X

Water

Crushed leaves or flowers

Cold water in

Purified plant oils

Water

Heat

Heat

▲ The equipment needed for Miss Jackson's steam extraction process

Step 1: Crush the leaves or flowers of the plant with a pestle and mortar and place into a flask.

Step 2: Pass steam from boiling water into the flask, which carries away some of the perfume chemicals in the plant.

Step 3: The steam is then condensed into a liquid which contains separate layers of water and perfume oils.

1 What is the solvent in this method?
2 Suggest what temperature the water is at in the flask on the left of the diagram.
3 What is the name of the piece of apparatus which is labelled X on the diagram?
4 What separating technique does this method most remind you of: evaporation, filtration, distillation, chromatography?
5 The conical flask on the right of the diagram contains two liquids which do not mix, so they are in separate layers. The valuable perfume oils are in the top layer. Suggest how you would separate the oils from the water below.

▲ Forensic scientists help investigate evidence at crime scenes

The fraudulent will

Miss Jackson told her nephew that she was leaving her entire fortune to him, and she even showed him her will. However, when she died, her will had been changed! The new will left all of her fortune to her three nieces, in equal shares. Someone was guilty of changing her will, but who?

A sample of ink was taken from the handwritten will that had been forged. A sample of ink was taken from the pens belonging to each of the three nieces. The forensic scientists wanted to find out if the ink from the forged will matched the ink from one of the pens belonging to the three nieces. They used chromatography to compare the inks. Here is how they set up the chromatogram before and after running the experiment.

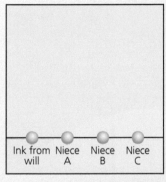

▲ The chromatogram before running the experiment

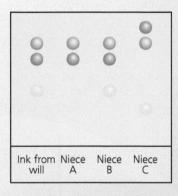

▲ The chromatogram after running the experiment

6 Why was it important that the horizontal line was drawn in pencil and not in ink?

7 Why would it have been a problem if the pencil line was not horizontal?

8 What would have happened if the solvent level in the beaker was above the pencil line?

9 What can you conclude from the chromatogram? Justify your conclusion in as much detail as you can using evidence from the chromatogram.

10 What **limitations** are there from using this evidence to convict one of the nieces of the crime?

Reactions

Learning objectives

11 Metals and non–metals

In this chapter you will learn...

Knowledge

- that metals react with oxygen to form oxides that are bases
- that non-metals react with oxygen to form oxides that are acids
- about the relative reactivity of metals and reactivity series
- that some metals react with acids to produce salts and hydrogen
- the definitions of the terms metal, non-metal, displacement, oxidation and reactivity

Application

- how to deduce a word equation to describe chemical reactions
- how to use particle diagrams to represent oxidation, displacement and metal–acid reactions
- how to place an unfamiliar metal into the reactivity series based on information about its reactions

Extension

- how to justify the use of specific metals and non-metals for different applications, using data provided

12 Acids and alkalis

In this chapter you will learn...

Knowledge

- that the pH scale can be used to describe acidic, neutral and alkaline solutions
- that acids and alkalis can be corrosive or irritant and about suitable safety precautions
- about the difference between the pH of strong and weak acids, and some examples of each
- that mixing an acid and alkali produces a neutralisation reaction, forming a salt and water
- the definitions of the terms pH, indicator, base and concentration

Application

- how to identify the best indicator to distinguish between solutions of different pH, using data provided
- how to use data and observations to determine the pH of a solution and explain what this shows
- how to explain how neutralisation reactions are used in a range of situations
- how to describe how to make a neutral solution from an acid and alkali

Extension

- how to deduce the hazards of different alkalis and acids using data about their concentration and pH
- how to estimate the pH of an acid based on information from reactions

Metals and non-metals

Your knowledge objectives

In this chapter you will learn:
- that metals react with oxygen to form oxides that are bases
- that non-metals react with oxygen to form oxides that are acids
- about the relative reactivity of metals using a reactivity series
- that some metals react with acids to produce salts and hydrogen
- the definitions of the terms metal, non-metal, displacement, oxidation and reactivity

See page 119 for the full learning objectives.

» Transition: Choosing a material

Have you ever wondered why coins are made from metal? And why we use metal cutlery for eating our food, but very small children use plastic cutlery? Some objects are made from metals and some objects are made from non-metals. The spoon in the picture is made from a metal called stainless steel.

▲ Different objects are made from different materials

Metals all have similar properties. A property is a word that describes what a substance is like or how it behaves.

Common error

Do not confuse a material with a property. Plastic is not a property – it is a material. Remember that 'cheap' isn't a property either.

What are metals used for and why?

▲ Saucepans can be made from steel because it has a **high melting point** and is a **good conductor of heat**

▲ Electrical wires are made from copper because it is an **excellent conductor of electricity** and it is **flexible**

What are non-metals used for and why?

▲ This bucket is made from plastic because it is **flexible**, **strong** and **waterproof**

▲ This window is made from glass because it is **transparent**, **waterproof** and **strong**

Worked example

The tyres on a car are made from rubber. The door of a car is made from steel. Explain why these materials are chosen for these uses, using ideas about their properties.

*Rubber is used for the tyres **because** it grips the road due to friction, which is important when braking, steering and accelerating. It is also flexible which helps to cushion the car from bumps in the road. Steel is used for the car door because it is strong, to protect the passengers in a crash.*

Apply >

1 Sort these properties into those which are typical of metals and those which are typical of non-metals.

Good conductor of heat

Bad conductor of electricity

Brittle (shatters when hit)

Malleable (can be beaten into a new shape)

Shiny

Dull

2 Knives and forks can be made of plastic as well as metal. What properties do they have in common?

3 Why are spoons for young children made from plastic instead of metal?

4 Radiators which are used to heat your house are made from steel. Use ideas about properties to explain why.

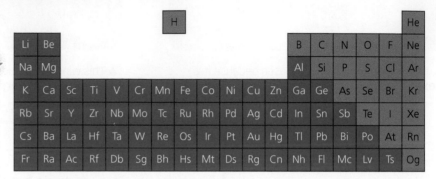

▲ Metals are often used in the construction of large structures like bridges and buildings

» Core: Properties of metals and non-metals

Metals

There are around 120 chemical elements and most of them are **metals**. You will find them on the left and middle of the periodic table. You are going to learn more about the periodic table in Chapter 9 of Pupil's Book 2.

▲ On this diagram of the periodic table, the metals have been shaded red and the non-metals have been shaded blue

Some metals are not pure elements. These are called **alloys**. Brass is an example of an alloy. It is a mixture of copper and zinc. Steel is also an alloy. Alloys are not listed on the periodic table because they are not pure elements.

Metals have a lot of properties in common. This means that they behave in similar ways. Here are the typical properties of metals, together with an example of how this property is useful.

Property	Example
Good conductor of electricity	Copper is used for electrical wiring
Good conductor of heat	Steel is used for some saucepans
Malleable	Titanium is used to make aeroplanes
Ductile	Aluminium is used to make pylon wires
Shiny	Platinum is used for jewellery

Most metals have a high melting point, which means that they are solids at room temperature. However, mercury is a metal element which is a liquid at room temperature. Mercury is used in some thermometers because it expands a lot when it is heated.

Some metals are **magnetic**. The most common metal which is magnetic is iron (and steel, which is an alloy containing iron). Nickel and cobalt are also magnetic.

Key words

Metals are a group of substances that are all good conductors of electricity and heat. They are usually shiny, malleable and ductile, and almost all are solid at room temperature.

An **alloy** is a mixture of a metal with another element which is usually a metal.

A substance is **malleable** if it can be beaten into a new shape.

A substance is **ductile** if it can be drawn into wires.

▲ Mercury droplets from a mercury thermometer

Key fact

→ Metals and non-metals are separated on the periodic table by a staircase line.

Non-metals

The **non-metal** elements are shaded blue in the diagram of the periodic table on the page opposite. Some non-metals join to form compounds. Examples include water, carbon dioxide and methane. Non-metals often have the properties listed in the table below.

Property	Example
Poor conductor of electricity	Plastic is used to insulate electrical wires
Poor conductor of heat	Polymer fibres are used to make oven gloves
Brittle	Glass on fire alarm trigger point
Many non-metals have a dull appearance	The surface of paper can be written on easily

Non-metal elements are usually gases, but bromine is a liquid at room temperature. Non-metal compounds could be gases (e.g. carbon dioxide), liquids (e.g. alcohol) or solids (e.g. plastics).

Key words

Non-metals are elements, compounds or mixtures that are not metals. They can be solids, liquids or gases, and have a range of properties.

A substance is **brittle** if it shatters when it is hit with a hammer, or if it breaks when you try to bend it.

Worked example

Diamond is a pure form of the element carbon. It has a very high melting point and is a poor conductor of electricity. It is brittle. Suggest whether diamond is a metal or non-metal and explain your answer.

I think that diamond is a non-metal. This is because it is a poor conductor of electricity and it is brittle. The high melting point doesn't help me decide, because some non-metals and most metals have high melting points.

Know >

1 List three metal elements which are magnetic.

2 Why is steel not listed on the periodic table?

3 Which metal element is a liquid at room temperature?

4 Which non-metal element is a liquid at room temperature?

Apply >>

5 Iodine is a solid at room temperature and is a poor conductor of heat and electricity. Is it a metal or non-metal? Explain your answer.

6 A spoon used for cooking is made from stainless steel (a metal) but it has a plastic handle. Explain why, using ideas about properties.

Extend >>>

7 Look at the data table below and decide which metal would be the best to use for making pylon wires to carry electricity over large distances. Explain your answer using ideas about properties.

Metal	Relative conductivity of electricity	Relative conductivity of heat	Ductile?
A	80	25	Yes
B	85	50	No
C	60	70	Yes

Enquiry >>>>

8 Describe an investigation which will allow you to decide which metal has the highest electrical conductivity. How will you make it a fair test? What will your independent and dependent variables be? What equipment will you use?

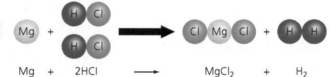

▲ Magnesium reacting with a dilute acid. The burning splint is used to test for hydrogen

» Core: Metals, acids and the reactivity series

Metal + acid

Some metals react with acids. When a metal reacts with an acid, bubbles are seen and eventually the metal disappears. The bubbles show that a gas is made and the test tube gets warm. The gas makes a 'squeaky pop' sound when a burning wooden splint is used to test the gas.

The general word equation for this reaction is:

Metal + acid → salt + hydrogen

This is easy to remember because the first letter of each of the chemicals spells the word MASH. The name of the **salt** produced depends on the metal used and the name of the acid. For example:

Magnesium + hydrochloric acid → magnesium chloride + hydrogen

$$Mg + 2HCl \longrightarrow MgCl_2 + H_2$$

▲ A particle diagram can help us to understand the reaction

Calcium + nitric acid → calcium nitrate + hydrogen

Zinc + sulfuric acid → zinc sulfate + hydrogen

Key fact

→ Metal + acid → salt + hydrogen

Making a reactivity series

Not all metals react with acids. Some react violently and some metals do not react with dilute acids at all.

Metal	Observations with dilute sulfuric acid
Zinc	Many small bubbles appear and rise to the surface
Copper	No visible reaction
Magnesium	Large bubbles are produced quickly; feels warm
Calcium	Large bubbles are produced very quickly; feels very hot
Silver	No visible reaction

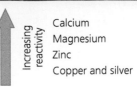

Key words

Reactivity is how easily a substance reacts with other chemicals.

A **reactivity series** shows which substances in a group are most and least reactive.

By making observations of these reactions, we can put the metals into a **reactivity series**. However, we cannot decide from these observations whether copper or silver is more reactive. To work this out, we would need to react those metals with another substance, like oxygen.

Increasing reactivity

Calcium
Magnesium
Zinc
Copper and silver

Worked example

Four metals (A, B, C and D) were reacted separately with dilute nitric acid. B reacted very slowly, producing small bubbles. A did not react at all. C reacted to produce small bubbles quickly. D reacted very quickly and produced large bubbles of gas. Place these metals into order of reactivity.

D was the most reactive, then C and then B. Metal A was the least reactive.

Know >

1 Which gas is made when a metal reacts with an acid?

2 Describe how to test for hydrogen gas. Include the positive test result in your answer.

3 Which acid reacts with magnesium to produce magnesium chloride?

Apply >>

4 When nitric acid reacts with zinc, a solution of zinc nitrate is made and hydrogen gas is produced. Write a word equation to describe this reaction.

5 Name the salt produced when aluminium reacts with hydrochloric acid.

6 Copy and complete this word equation:

Magnesium + sulfuric acid →

7 Copy and complete this word equation:

Calcium + hydrochloric acid →

Extend >>>

8 Which acid would you need to react with iron to produce iron chloride and hydrogen?

9 Work out the missing reactants for this chemical equation.

———— + ———— → zinc chloride + hydrogen

10 Work out the missing reactants for this chemical equation.

———— + ———— → magnesium nitrate + hydrogen

Enquiry >>>>

A student wants to place five metals into order of reactivity by observing their reactions with dilute hydrochloric acid.

11 What should the student try to keep the same during this investigation in order to make it a fair (valid) test?

12 What could the student do if two of the metals did not react with the acid?

13 What safety precautions should the student follow during this investigation?

» Core: How do metals and non-metals react with oxygen?

Metal + oxygen

When a metal reacts with oxygen, a metal oxide is formed. This is called an **oxidation reaction**. All metal oxides are solids. For example:

Magnesium + oxygen → magnesium oxide

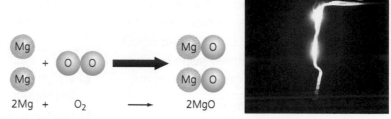

$$2Mg + O_2 \longrightarrow 2MgO$$

▲ The reaction of magnesium with oxygen to form magnesium oxide

Some metals react with oxygen very quickly. Other metals react slowly with oxygen, and some metals do not react with oxygen at all. We can use observations of these reactions to place metals into a reactivity series.

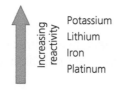

Increasing reactivity

Potassium
Lithium
Iron
Platinum

Metal	Reaction with oxygen
Potassium	Reacts very quickly. Surface goes grey-purple.
Platinum	Does not react with oxygen.
Lithium	Reacts quite quickly over a minute or so.
Iron	Slowly rusts over several months.

Non-metal + oxygen

Many non-metal elements react with oxygen in oxidation reactions. Sometimes the non-metal oxide is a gas (e.g. sulfur dioxide, carbon dioxide). Sometimes the non-metal oxide is a liquid (e.g. water). Sometimes the non-metal oxide is a solid (e.g. silicon oxide). Here is an example:

Sulfur + oxygen → sulfur dioxide

$$S + O_2 \longrightarrow SO_2$$

▲ The reaction of sulfur with oxygen to form sulfur dioxide

Acidic or alkaline oxides?

Many non-metal oxides react with water. If this happens, they usually make **acids**. Here are some examples.

Sulfur trioxide + water → sulfuric acid

Carbon dioxide + water → carbonic acid

Phosphorus oxide + water → phosphoric acid

Nitrogen dioxide + water → nitric acid

All metal oxides are bases. A **base** is a chemical which can neutralise an acid. If a base dissolves in water, it makes an **alkali**. Not all bases dissolve in water, but here are some that do…

Sodium oxide + water → sodium hydroxide

Calcium oxide + water → calcium hydroxide

Worked example

Copper does not react with oxygen at room temperature. It must be heated with a Bunsen flame and then it slowly turns black, forming copper oxide. When sodium is left exposed to the air, it reacts within ten seconds to form sodium oxide. Magnesium must be heated with a Bunsen flame before it reacts with oxygen in the air, but it burns quickly with a bright flame. Put these metals in order of increasing reactivity.

The least reactive is copper. Magnesium is more reactive. Sodium is the most reactive.

Know >

1 When iron is burned in air, it makes iron oxide. Suggest whether iron oxide is a solid, liquid or gas.
2 Carbon can be easily burned in air to make carbon dioxide. Is carbon dioxide a solid, liquid or gas?

Apply >>

3 Use ideas about platinum to explain why it is used for jewellery.

4 Suggest the pH of a solution made from dissolving sulfur dioxide into water.
5 Suggest the pH of a solution made when sodium oxide reacts with water.
6 Explain why a solution of sulfur dioxide and a solution of sodium oxide will react with each other.

Extend >>>

7 Name the products when a solution made from phosphorus oxide and water reacts with a solution made from calcium oxide and water.
8 Write a word equation to show what happens when a solution made from nitrogen dioxide and water reacts with a solution made from potassium oxide and water.

Enquiry >>>

A student decided to research the formation of oxides on the internet because her teacher told her that the school did not have any pure oxygen.

9 Why could the teacher still demonstrate the reaction of some metals and non-metals with oxygen, even though the school did not have an oxygen cylinder?
10 What advice would you give to the student when she was searching for this information online so that she could be confident that her research would be accurate and reliable?

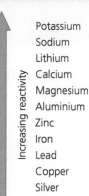

Increasing reactivity

Potassium
Sodium
Lithium
Calcium
Magnesium
Aluminium
Zinc
Iron
Lead
Copper
Silver
Gold

» Core: Displacement reactions

In a **displacement reaction**, a more reactive element will replace a less reactive element in a compound. We can use the reactivity series to make predictions about whether one metal will displace another metal in a reaction.

We can predict that magnesium will displace copper from copper chloride. But magnesium will not displace calcium from calcium chloride.

Magnesium + copper chloride → copper + magnesium chloride

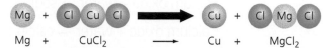

Mg + $CuCl_2$ ⟶ Cu + $MgCl_2$

▲ The displacement reaction between magnesium and copper chloride

Displacement reactions can sometimes happen very quickly and release a lot of energy, if there is a big difference between the reactivity of the two metals. The reaction between solid magnesium and solid silver nitrate can be started with one drop of water and is explosive, reacting at over 3000 °C! But the reaction between copper metal and a solution of silver nitrate takes over an hour. The colourless silver nitrate solution is slowly turned into blue copper sulfate solution. Crystals of silver metal grow on the wire spiral.

Copper + silver nitrate → silver + copper nitrate

▲ The reaction between copper wire and silver nitrate solution

We can also use observations of displacement reactions to place a new metal into the reactivity series. Tin metal is added to a solution of iron nitrate and no reaction occurs. But when tin is added to a solution of lead nitrate, a displacement reaction occurs:

Tin + lead nitrate → lead + tin nitrate

So we can conclude that tin must be more reactive than lead but less reactive than iron. So it must go between these two metals in the reactivity series.

Worked example

Predict what will happen when zinc is added to a solution of copper sulfate and also to a solution of potassium sulfate. Write word equations for any reactions that occur.

Zinc will react with copper sulfate and displace the copper because zinc is more reactive than copper. The reaction will be: zinc + copper sulfate → copper + zinc sulfate. There will be no reaction when zinc is added to potassium sulfate because the zinc is less reactive than the potassium.

Know >

1 Which metal is more reactive, gold or sodium?

2 What name is given to a reaction when a more reactive metal replaces a less reactive metal in a compound?

Apply >>

3 List three metals which can be displaced using calcium, and one which will not be displaced.

4 A student wanted to obtain pure lead metal from a solution of lead nitrate. She added copper metal to the lead nitrate solution. No reaction occurred. Explain why her experiment was not successful, and suggest what she should do to obtain lead from lead nitrate solution.

Extend >>>

5 Copy and complete the following equations. Write 'no reaction' after the arrow if no displacement reaction occurs.

Zinc + lead nitrate →

Copper + sodium chloride →

Iron + copper bromide →

Enquiry >>>>

A student is given three unknown metals (labelled J, K and L). He is also given samples of the three metal nitrates (J nitrate, K nitrate and L nitrate).

6 Describe a brief outline method he could follow to use displacement reactions to work out which metal is the most reactive and which is the least reactive.

7 Draw up a results table that he could use.

8 How could he interpret his results to decide which was the most reactive and the least reactive?

» Extend: Choosing a substance for a particular use

We can choose which material will be better for a particular use by evaluating data about the properties and cost of each material. First, you need to decide which properties are most important and least important for a particular use.

Key word

When you **evaluate** something, you consider the advantages and disadvantages and come to a decision which you can justify.

Common error

The cost of a substance is not a property. Properties describe what the material is like and what things it does well and badly. The cost of a metal depends on how much people want it and how much of it is available to buy.

Example – making the legs of a school chair

The metal used for the chair legs needs to be strong so it will support the weight of the student sitting on the chair. The metal needs to be malleable so that it can be easily made into a metal tube and then bent to form the shape needed for the chair legs. It would be good if it was resistant to corrosion because then the legs of the chair wouldn't need to be painted. It does not need to be a good conductor of heat as this is irrelevant for a chair! It also does not matter too much if the chair legs are heavy (dense) or light. Once you have considered the properties of the metals, you should also consider the cost of each metal.

Metal	Relative strength	Malleable	Resistance to corrosion	Relative conductivity of heat	Density (g/cm^3)	Cost per kg (£)
Aluminium	6	Yes	Highly resistant to corrosion	7	2.7	1.50
Steel	9	Yes	Quite resistant to corrosion	6	7.6	0.14
Copper	4	Yes	Highly resistant to corrosion	9	9.0	3.90
Lead	3	Yes	Highly resistant to corrosion	4	11.3	1.00

The best metal to make chair legs would probably be steel. This is because it is the strongest and this is the most important property because the chair needs to hold the weight of the person sitting on it. All of the metals are malleable, and the steel is only quite resistant to corrosion, so it would probably need to be painted. Steel is also the cheapest metal in the table.

Tasks

A sports car

1. List the properties which you think are important for the body panels of a cheap family car. The body panels are parts like the doors, bonnet, roof, etc. For each of the properties, explain why you think it is important.

2. Choose one of the metals from the table which you think will be the best for making the body panels of a sports car. Explain why you think that this is the best metal.

3. For the other three metals, explain why they are not as good for this use.

▲ Sports cars need to be safe as well as fast

Enquiry:
Making a reactivity series

You are given six metals and asked to place them into a reactivity series. You decide to observe their reactions with air, with acid and with each other (displacement reactions).

1. Describe how you would observe their reactions with air in order to obtain valid and useful results to help you decide which metals are more or less reactive.
2. Describe using a step by step method how you would observe their reactions with acid. Include details of the safety precautions you would take in this part of the reaction.
3. Describe how you would observe the displacement reactions of the six metals.
   Check your answer carefully to make sure you have communicated your ideas clearly and with correct spelling, punctuation and grammar.
   Design a results table that you could use.

Reactions with oxygen

The observations of the six metals with oxygen are detailed in the table below.

Metal	Observations
A	No reaction with air, even when heated strongly
B	Reacts when cold with air to give a dull grey surface within 15 s
C	Reacts with air when heated gently to give a pale grey surface
D	No reaction with air, even when heated strongly
E	Reacts when cold with air to give a pale grey surface within 3 minutes
F	No reaction with air, even when heated strongly

4. What can you conclude from these observations?
5. What can you *not* conclude from these observations?

Reactions with dilute acid

The six metals were then reacted with dilute acid and the results were as follows.

Metal	Observations
A	No reaction with the dilute acid.
B	Violent reaction with dilute acid. Sparks, flammable gas and small explosion.
C	Vigorous reaction with dilute acid. Large fast bubbles which catch on fire.
D	Lots of small bubbles. Reaction gives off heat.
E	Violent reaction with dilute acid. Sparks, flammable gas and small explosion.
F	Lots of very small bubbles are produced from the surface of the metal.

▲ One of the metals in this investigation reacting with dilute acid

6 What can you conclude about these observations?

7 Taking the observations from the first two experiments in this investigation, place the metals into order of reactivity.

The displacement reactions

The displacement reactions of the six metals were then investigated by another group of students. They didn't have a results table, but wrote down their observations.

> We used six metals, and solutions of the sulfates of all six metals. Metal A didn't do anything when we put it into the five sulfate solutions. We didn't bother testing A with a solution of A sulfate. Metal B reacted with all of the solutions we added it to. Sometimes it was a fast and hot reaction, like with F and A. Other times, metal B gave a slower reaction, like with C and E. Metal C had a reaction with solutions of D, A and F, but not with solutions of B and E. Metal D only displaced A and F from their solutions. Metal E didn't do anything with solution B, but metal E did react with the other four solutions. Metal F was pretty rubbish, and only reacted with the solution of A sulfate.

▲ One of the displacement reactions in this investigation

8 Using your results table from Question 3, indicate which combinations of metal and metal sulfate solution gave a displacement reaction. Use a tick or cross to show whether a reaction took place.

9 Do the results from the displacement reactions support your reactivity series from Question 6? If not, what is your final decision for a reactivity series of these six metals? Justify your final answer by referring to specific pieces of evidence.

10 Evaluate this approach to deducing a reactivity series. Can you suggest any improvements?

12 Acids and alkalis

Your knowledge objectives:

In this chapter you will learn:

- that the pH scale can be used to describe acidic, neutral and alkaline solutions
- that acids and alkalis can be corrosive or irritant and about suitable safety precautions
- about the difference between the pH of strong and weak acids, and some examples of each
- that mixing an acid and alkali produces a neutralisation reaction, forming a salt and water
- the definitions of the terms pH, indicator, base and concentration

See page 119 for the full learning objectives.

» Transition: What are acids and alkalis?

The sharp taste of vinegar is caused by an acid called ethanoic acid. Citrus fruits like lemons, limes and oranges also have a sharp taste, due to citric acid.

▲ Citrus fruits contain an acid called citric acid

Acids are a group of chemicals that all react in similar ways. For example, you may already have seen what happens when vinegar is added to sodium hydrogencarbonate (also called bicarbonate of soda). A chemical reaction happens and bubbles of gas are produced. If you use a different acid, like citric acid or hydrochloric acid, the reaction will be similar.

Alkalis are chemically opposite to acids. Acids can be neutralised by alkalis to produce a solution that is neither an acid or an alkali – it is neutral.

► The reaction between vinegar and sodium hydrogencarbonate produces bubbles of carbon dioxide gas

The pH scale

We measure how acidic or alkaline a solution is using the **pH scale**. The pH scale goes from 0 to 14, with pH 7 being in the middle, which is a neutral solution.

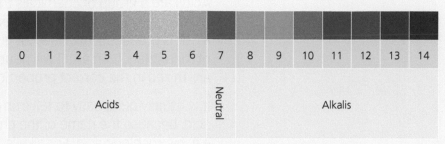

0	1	2	3	4	5	6	7	8	9	10	11	12	13	14

Acids   Neutral   Alkalis

▲ The pH scale is used to measure how acidic or alkaline a substance is

Worked example

A student decided to test the pH of some substances using a pH meter. Copy her results table below and fill in the gaps.

Substance	pH value	Acid, alkali, or neutral
Vinegar	4	
Milk	7	
Toothpaste	9	
Water	7	
Bleach	14	
Hydrochloric acid	0	

Substance	pH value	Acid, alkali, or neutral
Vinegar	4	Acid
Milk	7	Neutral
Toothpaste	9	Alkali
Water	7	Neutral
Bleach	14	Alkali
Hydrochloric acid	0	Acid

Remember that anything which is pH 7 is neutral; anything with a pH of less than 7 is acidic; and anything with a pH greater than 7 is alkaline.

Apply >

1 A student tests washing up liquid and finds that it has a pH which is 8. Is the washing up liquid acidic, neutral or alkaline?

2 Suggest a pH value for a sample of lemon juice.

3 Suggest what happens to the pH of vinegar when it reacts with sodium hydrogencarbonate.

» Core: The pH scale

Acids and alkalis

Acids are a group of chemicals that have a pH of less than 7. They all react in similar ways, and when they do react, a chemical called a salt is produced. Alkalis are a group of chemicals which have a pH of more than 7. An alkali will neutralise an acid if the two solutions are mixed in the correct proportions.

It is usually quite easy to tell from the name of a substance if it is an acid, because the name of the chemical contains the word 'acid', e.g. hydrochloric acid. For alkalis, it is a bit harder. Many common alkalis end in the word 'hydroxide', such as sodium hydroxide, or ammonium hydroxide.

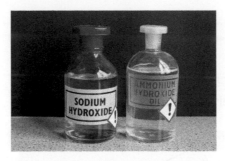

▲ Sodium hydroxide and ammonium hydroxide are common alkalis used in the science laboratory

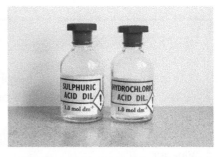

▲ Sulfuric acid and hydrochloric acid are common acids used in the science laboratory

Key word

A **neutral** solution has a pH of 7 and is neither acidic nor alkaline.

The pH scale shows how acidic or alkaline a solution is. A solution which is neither acidic nor alkaline is described as being **neutral**, and it has a pH of 7.

0	1	2	3	4	5	6	7	8	9	10	11	12	13	14

Strong acids | Weak acids | Neutral | Weak alkalis | Strong alkalis

▲ The pH scale shows how acidic or alkaline a solution is

Strong and weak acids

Some acids are strong acids. Strong acids have the lowest pH values, and this means that they react very vigorously with chemicals like metals and alkalis. Hydrochloric acid, sulfuric acid and nitric acid are all strong acids. Other acids are weak acids, and they do not react so vigorously. Ethanoic (acetic) acid and citric acid are weak acids. Strong acids are more likely to be dangerous than weak acids.

Alkalis can be strong or weak as well. Sodium hydroxide is a strong alkali, but ammonium hydroxide is a weak alkali.

Concentration

The concentration of a solution is how many solute particles (e.g. acid particles) are dissolved in one litre of solution. If you add water to a solution it becomes less concentrated, and more dilute (just like when you add water to concentrated orange squash). A strong acid (like hydrochloric acid) can be either concentrated or dilute. A weak acid (like ethanoic acid) can either be concentrated or dilute.

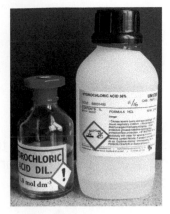

▲ Concentrated and dilute should not be confused with strong and weak. Hydrochloric acid is always a strong acid, even when it is dilute

Key facts

It is useful to know the formulae of some common acids…
→ HCl is hydrochloric acid
→ HNO_3 is nitric acid
→ H_2SO_4 is sulfuric acid

It is useful to know the formulae of some common alkalis…
→ NaOH is sodium hydroxide
→ KOH is potassium hydroxide
→ NH_4OH is ammonium hydroxide

Worked example

A solution is tested with a pH meter. Its pH is 8. What does this tell you?

The solution is weakly alkaline.

Know >

1 Name three strong acids.

2 How can you make an acid less concentrated?

3 What is the pH of a neutral solution?

Apply >>

4 Bleach is used in the kitchen and bathroom to kill bacteria. It has a pH of 14. What does this tell you?

5 Fresh milk has a pH of 7, but when it turns sour the pH decreases. What can you deduce from these facts?

6 Which element is present in all acids?

Extend >>>

7 Would you describe the pH scale as a continuous scale or a discontinuous scale? Explain your answer.

8 Is it possible to make a strong acid weaker? Explain your answer.

Enquiry >>>>

9 You are given some sulfuric acid which has a concentration of 2 mol/dm³. You need to dilute it so that concentration has been halved (1 mol/dm³). Describe how you would do this.

» Core: Acids reacting with alkalis

Hazards and safety precautions

Acids and alkalis can both be dangerous, depending on their strength and concentration. Chemists use hazard warning labels to show how dangerous a solution of an acid or alkali is, and you can then decide what safety precautions to take.

Concentrated strong acid, concentrated weak acid, concentrated strong alkali, concentrated weak alkali	DANGER: Corrosive Safety precautions: wear goggles and gloves	
Dilute strong acid (e.g. hydrochloric acid) or dilute strong alkali (e.g. sodium hydroxide)	WARNING: Irritant Safety precautions: wear safety glasses	
Dilute weak acid (e.g. ethanoic acid) or dilute weak alkali (e.g. ammonium hydroxide)	Low hazard	These solutions do not need a hazard warning label.

▲ A pH meter being used

Key word

An **indicator** is a chemical which can be used to identify whether a substance is acidic or alkaline because it changes colour.

Indicators

The colours we normally associate with different pH values are actually the colours that universal indicator changes when it is added to solutions with those pH values. Universal indicator is a **full range indicator**, because it can give us an approximate measurement of the pH. A more accurate way of measuring pH is to use a pH meter.

Indicators are chemicals that change colour depending on the pH of the solution they are added to. Most indicators only have two colours. Methyl orange is red in acids and yellow in alkalis. Phenolphthalein is colourless in acids and pink in alkalis.

Acid-alkali neutralisations

When an acid is neutralised by an acid, a salt and water are produced. The name of the salt depends on the name of the acid used and the name of the alkali used.

Acid + alkali → salt + water

For example:

Hydrochloric acid + sodium hydroxide → sodium chloride + water

HCl + NaOH ⟶ NaCl + H₂O

To make a neutral solution of a salt, you should follow the following steps.

1 Measure 20 cm³ of the alkali into a beaker, using a measuring cylinder.

2 Add two drops of universal indicator.

3 Measure 30 cm³ of the acid into a different beaker.

4 Add the acid slowly into the alkali (ideally using a dropper pipette) until the indicator shows that you have a neutral solution.

5 Measure the volume of the unused acid, and subtract that from the volume you started with (30 cm³) to calculate the volume of the acid that you added to the alkali.

6 Repeat the neutralisation using the same volumes of alkali and acid, but this time with no indicator.

Key facts

→ Acid + alkali → salt + water

Worked example

Write a word equation for the reaction between hydrochloric acid and potassium hydroxide.

Hydrochloric + potassium → potassium + water
acid   hydroxide   chloride

Know >

1 What colour does universal indicator go in a strong alkali?

2 What colour does universal indicator go in a weak acid?

3 A bottle of acid has the following label on it. What can you conclude about it?

Apply >>

4 State two safety precautions when working with concentrated alkalis.

5 Write a word equation for the reaction between hydrochloric acid and lithium hydroxide.

6 The table shows the colour of two different indicators at different pH values.

Indicator	pH 2	pH 4	pH 7	pH 10	pH 12
Methyl orange	Red	Orange	Yellow	Yellow	Yellow
Phenolphthalein	Colourless	Colourless	Colourless	Pink	Pink

If you are given two solutions, one of which is water and the other is an alkali, which would be the best indicator to distinguish between water and the alkali? Explain why.

Extend >>>

7 Name the acid and alkali which will produce ammonium chloride when they are reacted together.

8 Name the acid and alkali which will produce sodium sulfate when they are reacted together.

Enquiry >>>

9 The juice from a beetroot or a red cabbage can be used as an indicator to identify acids and alkalis, but you don't know that colour it will go at different pH values. Describe how you could work out whether a mystery solution is acidic or alkaline.

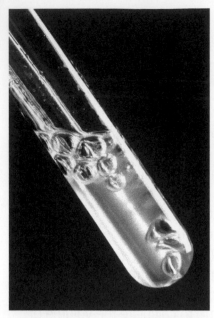

▲ Bubbling carbon dioxide through limewater makes it turn cloudy

» Core: Acids reacting with metal carbonates

Making carbon dioxide

When an acid is neutralised by a metal carbonate, there are three products: a salt, water and carbon dioxide. This means that bubbles will be seen during the reaction. Carbon dioxide can be identified because it turns limewater cloudy.

The general word equation for the reaction is:

Acid + metal carbonate → salt + water + carbon dioxide

For example:

Sulfuric + copper → copper + water + carbon
acid   carbonate   sulfate   dioxide

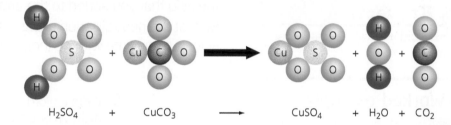

$$H_2SO_4 \quad + \quad CuCO_3 \quad \longrightarrow \quad CuSO_4 \quad + \quad H_2O \quad + \quad CO_2$$

> **Key facts**
>
> → Acid + metal carbonate→ salt + water + carbon dioxide

Naming salts

You can work out the name of any salt from the names of the acid and the base which have neutralised each other. The second part of the name of the salt comes from the acid. The first part of the name of the salt comes from the other chemical (the metal, metal oxide, metal hydroxide or metal carbonate).

For example:

Nitric + Potassium → Potassium + Water
acid   hydroxide   nitrate

$$HNO_3 \quad + \quad KOH \quad \longrightarrow \quad KNO_3 \quad + \quad H_2O$$

You can see that the water is made from the hydrogen part of the acid and the hydroxide part of the alkali.

Name of acid	Salt ends in...	Example
Hydrochloric acid	… chloride	Copper chloride
Nitric acid	… nitrate	Sodium nitrate
Sulfuric acid	… sulfate	Magnesium sulfate
Ethanoic acid	… ethanoate	Potassium ethanoate

Key facts

→ To work out the other products as well as the salt, you need to remember the general word equations we have met so far.
Metal + acid → salt + hydrogen (MASH equation)
Metal oxide + acid → salt + water
Metal hydroxide + acid → salt + water
Metal carbonate + acid → salt + water + carbon dioxide

Worked example

Write a word equation for the reaction between zinc and sulfuric acid.

Zinc + sulfuric → zinc + hydrogen
       acid   sulfate

Know >

1 What gas is made when an acid reacts with a metal carbonate?

2 Describe how to test for this gas.

3 What word will be the ending of the name of a salt made using nitric acid?

Apply >>

Copy and complete the following word equations.

4 Magnesium oxide + nitric acid →

5 Sodium carbonate + sulfuric acid →

6 Aluminium + hydrochloric acid →

Extend >>>

Copy and complete the following word equations.

7 —— + copper oxide → copper chloride + water

8 —— + —— → zinc nitrate + hydrogen

9 —— + —— → potassium + water + carbon
                  sulfate   dioxide

Enquiry >>>

You are given three white powders. One of them is sodium carbonate, another is magnesium oxide and the third is silicon oxide. Silicon oxide is insoluble and unreactive.

10 Describe how you could use a dilute acid to identify which of the solids is sodium carbonate.

11 Describe the safety precautions you would use when using a dilute acid.

12 Suggest how you could use the dilute acid to tell which of the two remaining was magnesium oxide.

» Extend: Making predictions about acids and alkalis

We can estimate the pH of an acid or alkali by using information about the way that it reacts.

▲ Using magnesium to test unknown solutions

A student was given samples of four different colourless solutions (A, B, C, and D). She did five tests with each solution. Here are the tests she did.

Test 1: What happens when I put a piece of magnesium metal into the solution?

Test 2: What happens when I put some marble chips (calcium carbonate) into the solution?

Test 3: What happens when I add some of the solution to a test tube which contains dilute sodium hydroxide (an alkali) and some universal indicator?

Test 4: What happens when I warm the solution in a beaker and then add some copper oxide powder and stir?

Test 5: What happens when I add some of the solution to a test tube which contains dilute hydrochloric acid and some universal indicator?

Here are her results:

| Solution | Observations with each test | | | | |
	Test 1	Test 2	Test 3	Test 4	Test 5
A	No reaction	No reaction	No change – stays purple	No reaction	Goes from red to purple
B	No reaction	No reaction	No change – stays purple	No reaction	No reaction
C	Lots of large bubbles	Lots of small bubbles	Goes from purple to red	Copper oxide disappears and solution turns blue	No change – stays red
D	Some small bubbles	Some small bubbles	Goes from purple to green	Solution turns very pale blue, some copper oxide powder remains	No change – stays red

From these results, we can identify whether each solution is acidic, alkaline, or neutral. Acids react with metals (Test 1), metal carbonates (Test 2), metal hydroxides (Test 3) and metal oxides (Test 4). So we can tell that solution C and solution D are both acidic. But the fact that solution C reacted more vigorously in these tests suggests that it is more acidic, so its pH must be lower. It would be sensible to suggest that the pH of solution C is 0, 1 or 2; and the pH of solution D could be 4 or 5.

Test 5 can be used to identify an alkali, as the universal indicator will be red to begin with, but adding a solution which is alkaline will neutralise the acid and make it alkaline. So solution A must be an alkali, perhaps pH 12, 13 or 14. Solution B doesn't react with any of the chemicals used in the tests, so we could suggest that it has a pH of 7.

Tasks

Three bottles of ethanoic acid (a weak acid) have been left in a storage cupboard over the school holiday and their labels have fallen off. The technician tells you that one of them is 'very dilute' with a concentration of $0.1\,mol/dm^3$, another is 'dilute' with a concentration of $2\,mol/dm^3$, and the third is 'concentrated ethanoic acid' with a concentration of $17\,mol/dm^3$.

1. What safety precautions would you follow when experimenting with these unknown acids?
2. How could you use a pH meter to decide which acid was which? Describe how you would interpret your results.
3. How could you use universal indicator to decide which acid was which? How would you interpret your results?
4. How could you use chemical reactions to decide which acid was which? Say which chemical(s) you would react them all with, and what you would expect to see for each acid.

Enquiry:
Which is the best indigestion remedy?

The lining of the stomach secretes hydrochloric acid to help digestion. The acid helps to digest proteins, by activating protease **enzymes** which work best when they are at a very low pH. The lining of the stomach also produces a special mucus to protect itself from being digested by its own acid!

Sometimes, the stomach produces too much acid, and this can rise into the oesophagus. Because the oesophagus is not protected by as much mucus as the stomach is, the acid attacks the lining of the oesophagus and causes pain, which is sometimes called heartburn, or acid indigestion.

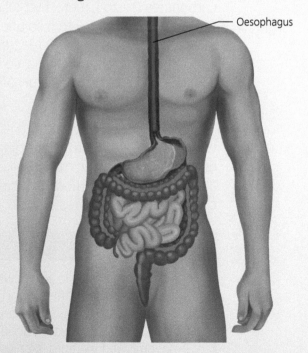

Oesophagus

▲ Acid from the stomach sometimes rises into the oesophagus and causes indigestion

Patients who suffer from heartburn can take indigestion remedies which neutralise some of the acid in the stomach. This relieves some of the discomfort. Many indigestion remedies contain metal carbonates.

▲ Antacid tablets can be used to treat indigestion

❶ What is the general word equation for the reaction of a metal carbonate with an acid?

❷ Write a word equation for the reaction between calcium carbonate and the specific type of acid found in the stomach.

❸ Patients tend to burp a little bit after taking an indigestion remedy. Suggest why, using the word equations you have written to help.

❹ Suggest why it might be dangerous to use a metal powder in an indigestion remedy, even though this would react with stomach acid and neutralise it.

A group of students wanted to test three indigestion remedies that they bought from the supermarket to see which one was the most effective in neutralising stomach acid. They decided to put some dilute hydrochloric acid into three boiling tubes and then they crushed up one of each type of tablet and added it to one of the boiling tubes. They recorded which one neutralised the acid the quickest.

❺ Why did the students decide to use dilute hydrochloric acid instead of stomach acid?

❻ What is the independent variable in this investigation?

❼ What is the dependent variable in this investigation?

❽ What variables must be controlled in order to collect valid results?

Here are the students' results.

Remedy name	Time taken for bubbles to stop (s)				Observations
	1	2	3	Mean	
Acideze	34	44	39		Bubbles produced quickly. No solid left at the end.
Kool Down	54	58	112		Lots of small bubbles. Some solid left at the end.
Neutraliser	16	20	21		Bubbles produced very quickly. No solid left at the end of the reaction.

Key words

The **range** of a data set is the difference between the smallest and largest measurement.

An **outlier** is a piece of data that does not fit the pattern.

The **mean** of a set of data is calculated by adding together the values and then dividing by the number of values which were added.

A set of results is **repeatable** if the repeat readings are close together. This means the range is small.

There is a **real difference** between two means if the range of the data sets does not overlap much.

❾ One of the times recorded for Kool Down is an **outlier**. Identify which one.

❿ Calculate the **mean** (average) reaction times. For the Kool Down data, ignore the outlier when calculating the average.

⓫ One of the students said that the results for Neutraliser were more **repeatable** than the data for Acideze. Why is this true?

⓬ Which reaction was the fastest?

⓭ One of the students suggested that Kool Down was the best antacid because it could neutralise more stomach acid than the other tablets, even if it wasn't the fastest. Justify this statement using data from the results table.

⓮ Calculate the **range** for Acideze and Neutraliser?

⓯ Why is a bar chart the most appropriate way to represent the data from the table?

⓰ Plot a bar chart of the mean reaction time for each type of remedy.

⓱ There is a **real difference** between these three sets of data. Explain why.

Earth

Learning objectives

13 Earth structure

In this chapter you will learn...

Knowledge
- about the processes in the rock cycle and how they convert rock from one type to another
- about the structure of the Earth
- the definitions of the terms rock cycle, weathering, erosion, mineral, sedimentary rock, igneous rock, metamorphic rock and strata

Application
- how to explain why a rock has a particular property based on how it was formed
- how to identify the causes of weathering and erosion and describe how they occur

Extension
- how to predict planetary conditions from descriptions of rocks on other planets

14 Universe

In this chapter you will learn...

Knowledge
- about the structure of the solar system
- that within the solar system, light spreads out from the Sun and is reflected by planets and moons
- to use appropriate models to explain day length, year length and the seasons
- that our solar system is a tiny part of a galaxy, one of many billions in the universe
- that light takes minutes to reach Earth from the Sun, four years from our nearest star and billions of years from other galaxies
- the definitions of the terms galaxy, light year, stars, orbit and exoplanet

Application
- how to describe the phases of the Moon and explain how they are caused using an appropriate diagram
- how to explain why places on the Earth experience different daylight hours and amounts of sunlight during the year
- how to describe how space exploration and observations of stars are affected by the scale of the universe
- how to explain the choice of particular units for measuring distance

Extension
- how to predict patterns in day length, the Sun's intensity or an object's shadow at different latitudes

» Transition: Rocks

We can see many different rocks in the world around us, and they are not all the same. Some rocks are different in their colour, or their texture. Some are made from crystals and some are made from layers which might be different colours or shades. Some rocks are very hard, whilst others are so soft you can scratch them with your fingernail.

Key words

Sedimentary rocks are formed when small particles settle out from slow moving water and then get squashed and glued together.

Fossils are the preserved remains or traces of living things from thousands or millions of years ago.

▲ The Mount Rushmore National Memorial in the USA was carved into very hard granite cliffs so that it will last a very long time before being eroded by weather

▲ Chalk is a type of rock which is so soft that it can be used to write on many surfaces

▲ Fossils are only found in sedimentary rocks, like these ones

Sedimentary rocks

Rocks which have layers are called **sedimentary rocks**. Sedimentary rocks were made when tiny particles settled out of slow moving rivers and oceans. They may contain **fossils**, which were formed over millions of years as minerals replaced the bones of dead animals which were trapped in the sediments as the rock was formed.

Igneous rocks

Rocks which are made from molten rock (hot liquid rock) are called **igneous rocks**. Igneous rocks are made from crystals which can often be seen with the eye.

▲ Granite is a very hard igneous rock which is often used for kitchen worksurfaces

Metamorphic rocks

Some rocks are made from other rocks which are heated and compressed under the surface of the Earth. These are called **metamorphic rocks**.

Worked example

Look at the information in the table below. Which of these rocks are sedimentary and which are igneous?

Rock name	Description
Limestone	Soft and made of layers which sometimes contain fossils.
Basalt	Made of microscopic crystals. Very hard.
Shale	Soft rock that breaks easily into layers.
Sandstone	Made of layers of sand grains glued together.
Gabbro	Made of large crystals. Very hard.

The sedimentary rocks are limestone, shale and sandstone because they are made from layers. Limestone also contains fossils. Basalt and gabbro are igneous rocks because they are made from crystals.

Apply »

1 A student finds a fossil in a rock at the beach. What kind of rock was it?

2 Look at this image of a rock. What kind of rock is it? Justify your answer.

» Core: The structure of the Earth

What is the Earth made from?

Have you ever wondered what it would be like to dig a hole all the way to the centre of the Earth? The Kola Superdeep Borehole in Russia was drilled between 1970 and 1992 and reached a depth of over 12 km. But the average radius of the Earth is 6371 km, so we have barely scratched the surface of the Earth.

▲ The Kola Superdeep Borehole in Russia

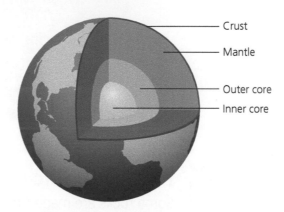

▲ The structure of the Earth

The Earth is generally described as having three layers: the core, the mantle and the crust. The crust is the solid outer part of the Earth. It is made from different rock types. The lowest parts of the crust are covered with water which makes the oceans. The crust is very thin in comparison with the diameter of the Earth.

The layer underneath the crust is called the mantle. The mantle is made from semi-solid rock which can flow very slowly over a long period of time. Underneath the mantle is the core, which scientists believe is made from iron and nickel. The outer core is a liquid, whilst the inner core is a solid.

How do we know this?

Studying the crust is relatively straightforward, because we can dig up the **rocks** and **minerals** that it is made from.

However, studying the mantle and the core is much harder because we have never drilled a hole through the crust and into these layers. Scientists use evidence about the way that **seismic waves** from earthquakes and explosions travel through the Earth to make deductions about the layers that we cannot see.

Common error

People often think that the mantle is liquid rock. The only liquid layer in the Earth is the outer core.

Key words

Rocks are mixtures of minerals found in the Earth's crust.

Minerals are solid chemical compounds that are found in the Earth's crust. Because each mineral is a compound, it will have a specific chemical formula.

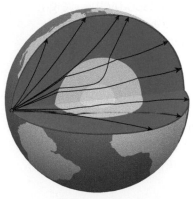

▲ Studying how seismic waves travel through the mantle and the core allows scientists to suggest what the Earth is made from

Worked example

Why is it impossible to write a chemical symbol equation for the reaction of chalk rock with hydrochloric acid?

Chalk is a rock, so it is a mixture of chemical compounds called minerals. Each mineral has a definite chemical formula, but the rock doesn't because it is a mixture.

Know >

1 Name the outermost layer of the Earth.

2 What is the only liquid layer of the Earth?

3 What is the mantle made from?

Apply »

4 Silicon oxide has the formula SiO_2 and it is one of the most abundant solids on Earth. Is it a rock or a mineral? Explain your answer.

5 If it was possible to dig a hole through the crust, what sorts of conditions would you experience inside the mantle?

Extend »»

6 Density is a measurement of mass per unit volume. What can you deduce about the relative densities of water, the crust, the mantle, and the core? Put them in order of density.

Enquiry »»»

7 How certain can scientists be that the core of the Earth is made from iron and nickel?

8 See if you can find out about some of the earlier scientific theories about the structure of the Earth. Why were they disproved in favour of newer ideas?

» Core: Types of rock

Classifying rock types

Scientists find it useful to classify things to help them to make sense of them, study their similarities and differences and make predictions about them. There is a huge range of rocks in the Earth's crust, but we group them into three types.

Sedimentary rocks

Sedimentary rocks are made from small particles which settle out from slow moving water in rivers, lakes or oceans. Over time, the sediment is glued together and compressed by the weight of sediment above it. Because the sediment is deposited on the bottom of the river, lake or ocean in layers, the sedimentary rock often has visible layers in it (called **strata**). Examples of sedimentary rocks include **limestone, chalk** and **sandstone**.

Fossils are sometimes found in sedimentary rock (but not in other types of rock). Fossils are formed when the hard parts of an animal (e.g. its shell or skeleton) are trapped in sediment when it turns to rock. Over time, the shell or sediment is replaced with minerals, which become the fossil.

▲ Fossils are only found in sedimentary rocks

Igneous rocks

Igneous rocks are formed from molten rock. Molten rock which is underground is called **magma** but if it reaches the surface, it is called **lava**. Igneous rocks can be formed from magma or lava. Igneous rocks are made from crystals, which are often visible to the naked eye, especially if the rock cooled down slowly underground. Examples of igneous rocks include **granite, basalt** and **obsidian**.

Metamorphic rocks

Metamorphic rocks are formed from existing rocks which are exposed to **heat** and **pressure** over a long period of time. They can be formed from sedimentary rocks or igneous rocks. Examples of metamorphic rock include **marble**, **slate** and **schist**.

▲ Granite has large crystals

▲ The Taj Mahal in India is made from marble, which is a metamorphic rock

Worked example

Mudstone is a sedimentary rock. Describe how it was formed and suggest what it will look like.

Mudstone was formed when small particles (sediment) settled out from slow moving water. It is likely to be made from visible layers because of the way that different sediments settled out over different periods of time.

Know >

1 What type of rock is limestone?

2 Name two igneous rocks.

3 Which rock type is made from layers of tiny particles compressed and stuck together?

Apply >>

4 Gabbro is a hard and dark rock which is made from large crystals. Suggest how it was made.

5 Conglomerate is a softer rock which is made from layers of rounded pebbles embedded in sand particles and clay. Suggest how it was made.

Extend >>>

6 Explain why fossils are never found in igneous rocks and why they are very unlikely to be found in metamorphic rocks.

Enquiry >>>

Hardness is a property which measures how easily a mineral or rock can be scratched. It is measured on the Mohs scale of hardness, with 10 being the hardest mineral (diamond) and 1 being the softest (talc).

7 You are given five minerals and need to decide on their order of hardness. Describe how you would do this.

8 How would you interpret the results?

» Core: The rock cycle

Weathering and erosion

Over very long periods of time (hundreds, thousands and millions of years), rocks at the surface of the Earth can be broken down by natural processes. This is called **weathering**.

Biological weathering is caused by living organisms. For example, a tree may grow in a crack in some rocks and its roots can force the crack to widen and a piece of rock to fall off. Chemical weathering is caused by chemical reactions which wear rock away. For example, rainwater is naturally acidic and can dissolve limestone to make limestone caves. Physical weathering doesn't involve chemicals or living things. For example, ocean waves can slowly wear away cliffs at the coast.

- Biological weathering

- Limestone caverns caused by chemical weathering

- A rock arch created by physical weathering

Erosion is when the smaller rocks and tiny particles caused by weathering are transported by moving water, ice or wind. Over time, huge amounts of rock can be moved by erosion.

The rock cycle

Over millions of years, rocks are converted from one type to another by very slow physical, chemical and biological processes which together make up the **rock cycle**. These processes are summarised in the diagram below.

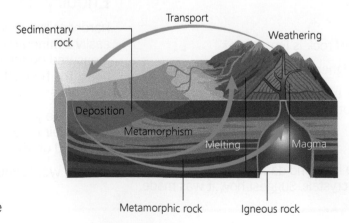

▶ A simplified diagram of the rock cycle

Rocks of any type can be weathered and eroded. The small particles of rock are transported down rivers to the ocean, where they settle out. Over time, they are turned into sedimentary rock. Movements of the plates that make up the Earth's crust can move sedimentary rocks upwards to make new land. Sedimentary rock can be heated and compressed underground to make metamorphic rock, or fully melted so that it turns into magma and then cools to form igneous rock.

Worked example

A glacier is a huge mass of ice which moves slowly downhill in a mountainous region. As it does so, it carves a deep and wide valley, carrying rocks with it. What type of weathering is this? Explain your answer.

This is an example of physical weathering because no living organisms are responsible for wearing the rocks away, and no chemical reactions are involved.

Know >

1 What type of weathering is caused by the action of waves on rocks at the coast?

2 What is the name of the process when rock is moved from one place to another by natural processes?

Apply »

3 When animals burrow in ground near to soft sedimentary rocks they can sometimes wear them away over time. Identify the type of weathering taking place.

4 Look at the picture of a rock structure found in a windy desert. Suggest how this rock formation was formed.

▲ Mushroom rock

Extend »»

5 Suggest which of the processes described in this section are the fastest, and which are the slowest.

6 Ceramics are non-metallic man-made materials which include the porcelain china used for making plates and bowls. Ceramics are usually made using heat to transform the properties of the starting material. How are ceramics similar to rocks, and how are they different from rocks?

» Extend: Rocks on other planets

Some of the rocks that geologists have studied are very old indeed, almost as old as the Earth itself. The Earth was formed approximately 4.5 billion years ago, at a similar time as the Sun and the other planets in the solar system.

As the Earth formed from dust and gas, it became very hot, so no chemicals were present as solids. The first rocks formed on the edge of the Earth as it began to cool down. The oldest rocks on Earth are estimated to be around 4.4 billion years old, and were reported in 2014, having been discovered in the Jack Hills in Australia.

The crust of the Earth is made from huge (continent-sized) plates of solid rock called tectonic plates. These plates are dragged around very slowly by movements of the semi-solid mantle underneath them. The plates move at a speed of a few centimetres per year, which is around the speed that your fingernails grow. These movements help to drive some of the processes in the rock cycle, such as melting old rock and forming new igneous rock in volcanoes.

Not all of the planets in the solar system have rocks on their surfaces. The table below summarises what scientists know (and think) about the composition of the eight planets in the solar system.

Planet	Composition
Mercury	Large dense iron core. Surface is made from very old igneous rocks. Lots of craters on the surface.
Venus	The internal structure is probably similar to Earth. Under the dense and hostile atmosphere of carbon dioxide and sulfur dioxide, there is a dry desert with slabs of angular igneous rock. There are still some volcanic eruptions. No rainfall reaches ground. There are some craters, which are in pristine condition. However, there are no small craters.
Earth	See page 150.
Mars	Dense metallic core. The crust contains similar compounds to Earth, but with more potassium. All of the rock on the surface is igneous. There are permanent ice caps, made from 70% solid water and 30% solid carbon dioxide. Geological activity is still taking place. There is currently no liquid water due to the low atmospheric pressure, but evidence of erosion channels or glaciers suggest that liquid water may have existed at some point. No recent weathering. Some impact craters, but far fewer than the Moon. Red colour of the dust on Mars is from iron oxide.
Jupiter	Made almost entirely of hydrogen and helium gases, although there may be some solid rocky material near the core. We don't really know!
Saturn	Mode mostly of hydrogen and helium gases, but there might be a solid rocky core. Scientists are not sure.
Uranus	The atmosphere of this giant planet is mostly hydrogen and helium. The solid part of the planet is mostly frozen water, ammonia and methane, with a small metallic rocky core.
Neptune	Similar to Uranus.

Task

① What can you suggest about geological processes (e.g. tectonic plate movement and formation of new igneous rock) in and around the Jack Hills region of Australia? Explain your answer.

Although it is not a planet, the Earth's Moon has a surface which is made from rock and dust. There are lots of craters on the surface. Footsteps and tyre tracks from every human landing on the Moon are still there. This is because there is no atmosphere, so there is no wind or rain to cause weathering or erosion. Analysis of the chemicals that make up Moon rocks show that they are made from the same chemicals found in the Earth's crust. This is consistent with the idea that the Moon was formed from a part of the Earth that was knocked off the young Earth by a huge collision.

Tasks

② There is one type of rock which is present on Earth but no other planet in the solar system. What type of rock is it, and why is it not present on any other planet?

③ Mercury and Venus both have relatively young igneous rocks on their surface, but the rocks on Mercury are much older. Why is this?

④ Earth has very few craters on its surface. Suggest some reasons why.

⑤ Mercury and Venus both have a lot of craters on their surfaces. The craters on Mercury have a range of sizes, from very small to very large. The craters on Venus are all above a certain size. Suggest why there are no small craters on Venus.

Enquiry:
Modelling stages in the rock cycle

Here are some practical activities to try in school (or at home in some cases) to model some of the processes involved in the rock cycle.

Sugar cubes in a plastic box

- Place a few sugar cubes into a plastic lunchbox.
- Shake the lunchbox gently ten times. Look inside and record your observations.
- Shake again ten more times. Look inside. Repeat these steps.

1. What process in the rock cycle does this represent?
2. What would happen if you tried this with pieces of chalk? Test your prediction by trying it if you can.
3. What would happen if you tried this with pieces of limestone? Test your prediction by trying it if you can.

Particles in a bottle

- Take an empty 500 cm³ bottle and add 1 cm depth of clean (washed) sand to the bottom.
- Now add 1 cm depth of small/fine clean gravel.
- Now add 1 cm depth of calcium carbonate powder if you can.
- Fill the rest of the bottle with water.
- Turn the bottle upside down many times until it is thoroughly mixed. Now leave it to settle out.

4. What process in the rock cycle does this represent?
5. What characteristic property of one specific type of rock can you see in the bottle once it has settled out?

Igneous intrusion

This is one to try at school.

- Put around 20 g of red candle wax into the bottom of a 250 cm³ beaker.
- Heat it gently over a Bunsen burner, tripod and gauze.
- Allow it to cool and freeze (solidify).
- Pour on a 1 cm layer of clean washed sand.
- Add a 2 cm layer of water on top of the sand.

▲ An igneous intrusion into the surrounding rock

- Now heat gently from underneath using a Bunsen/tripod/gauze with the gas tap turned half on/off. The air hole should be only open just enough to give you a clean blue flame.

- As the wax melts, its density decreases. When it gets less dense than the sand and water above it, it rises and pushes through the more dense layers which are above it.

- You have made an igneous intrusion!

6 What kind of rock does the sand represent?
7 The wax represents an igneous rock. What appearance would it have?

Moving tectonic plates

- You will need to find a very large beaker (1 litre or 2 litre).

- Pour in golden syrup until it is around 4 cm deep.

- Put the beaker into the refrigerator for around 24 hours.

- The syrup will now be a semi-solid, just like the mantle. You can prove this by turning the beaker upside down above your head!

- Snap a biscuit in half and place the two pieces touching each other on the surface of the syrup in the middle of the beaker.

- Place the beaker directly on a tripod.

- Heat very gently from underneath, using a Bunsen burner, under the middle of the beaker. Turn the gas tap down so that the flame is very small. Have the air hole open very slightly so that the flame is blue (clean) but not roaring.

- It may take 10–30 minutes for you to notice anything happening to the two pieces of the biscuit. Perhaps you can make a time-lapse movie using a mobile phone or tablet?

8 What happens to the two pieces of the biscuit?
9 Explain what has caused this to happen, using ideas about heat transfer and density.
10 Why is it important that the beaker is heated very slowly?

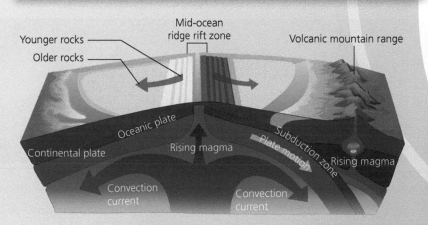

► Tectonic plates are forced apart by movements in the mantle beneath them

14 Universe

» Transition: Day and night

Your knowledge objectives:

In this chapter you will learn:
- about the structure of the solar system
- that within the solar system, light spreads out from the Sun and is reflected by planets and moons
- to use appropriate models to explain day length, year length and the seasons
- that our solar system is a tiny part of a galaxy, one of many billions in the universe
- that light takes minutes to reach Earth from the Sun, four years from our nearest star and billions of years from other galaxies
- the definitions of the terms galaxy, light year, stars, orbit and exoplanet

See page 147 for the full learning objectives.

When we look up into the sky we can see that the Sun appears to move across the sky during the day. Hundreds of years ago, people thought that this was because the Sun orbited (circled around) the Earth. But we now know that the Earth, and the other seven planets in the solar system all orbit the Sun. The reason that the Sun appears to move across the sky is that the Earth spins once every 24 hours. When it is day time, the part of the Earth you are on faces the Sun. When it is night time, your part of the Earth is facing away from the Sun.

The solar system

The Sun is a star. It is at the centre of the solar system. Eight planets **orbit** the Sun. The order of the eight planets can be remembered using a mnemonic like this:

Mercury	My
Venus	Very
Earth	Excellent
Mars	Mother
Jupiter	Just
Saturn	Served
Uranus	Us
Neptune	Noodles

Key word

To **orbit** an object in space means to circle around it, due to the attraction of gravity.

Common error

Many people still think Pluto is a planet. However, it was reclassified in 2006 as dwarf planet.

▲ The Sun and the eight planets of the solar system. The sizes of the Sun and planets are to scale, but the distances between the planets are not to scale (the planets are *much* more spread out than this)

Key word

A **celestial body** (or astronomical body) is an object in space. Planets, moons, stars, asteroids and comets are all examples.

Common error

Some people think that the Moon gives off light. In fact, it only reflects light from the Sun. The Sun is the only celestial object in the solar system that gives out light.

The Earth and the Moon

The Earth is the third planet in the solar system. It orbits the Sun once every year (365.25 days). The Moon is a **celestial body** which orbits the Earth. Earth has one moon, but some planets in the solar system have no moons and others have many moons.

▲ A full moon viewed from Earth

Worked example

Explain what causes day and night on Earth.

Day and night are caused by the rotation of the Earth. The Earth spins once every 24 hours. During the day time, the part of the Earth you are on is facing the Sun. When the Earth rotates and your part of the Earth faces away from the Sun, the Sun is not visible in the sky and it is night time.

Apply >

1 Venus rotates on its axis much more slowly than Earth does. What does this tell you about the length of a day on Venus?

2 How long does it take for Earth to complete one rotation about its axis?

3 How many moons does the Earth have?

» Core: The Earth in space

A year and a day

The Earth orbits the Sun, which is the **star** at the centre of the solar system. The Sun is the only thing in our solar system which produces light. In the night sky we can often see some of the other planets and the moon, but these are simply reflecting light from the Sun. Of course, other stars that we can see are producing their own light, but they are very far away – far outside our own solar system.

The time that it takes for the Earth to complete one **orbit** of the Sun is called a year (365.25 days). Whilst it is moving through space around the Sun, the Earth also spins on an axis which is 23° away from the vertical when compared with the direction it travels around the Sun. The time it takes for the Earth to spin once on its axis is called a day. On Earth, the Sun rises in the East and sets in the West due to the direction that the Earth spins.

The seasons

We know that the day length and the average temperature on Earth change during the annual cycle of the seasons. This is caused by the fact that the axis of the Earth's rotation is not exactly vertical.

Key words

A **star** is a celestial body which gives out light. A star may have a solar system of planets.

The **orbit** of a planet, moon or satellite is the path that it takes around a larger body.

▲ Looking west towards a sunset

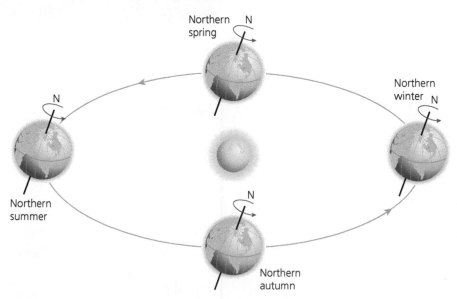

▲ Diagram showing how Earth's seasons are caused by the axis of rotation being tilted away from vertical

In the diagram you can see that someone living in the northern hemisphere (like the UK) has their summer when the north pole of the Earth is angled towards the Sun (on the left of the diagram).

This means that the light and heat from the Sun strike the surface of the Earth more directly, so the weather is warmer, the days are longer, and at midday the Sun is seen very high in the sky.

Six months later, the Earth has moved to the position on the right side of the diagram, so the countries in the northern hemisphere are now angled away from the Sun. This means that the day length is shorter, the weather is colder and the Sun stays quite low in the sky, even at midday. This is winter in the northern hemisphere, but of course at this time the southern hemisphere will be experiencing its summer, because it is facing the Sun more directly.

Worked example

New Zealand is in the southern hemisphere. It experiences its hottest weather and longest days during the months of December and January. Explain why.

The Earth rotates on an axis which is tilted relative to the orbit that it follows around the Sun. During December, the southern hemisphere of the Earth is angled towards the Sun, so the energy from the Sun hits the Earth's surface directly, leading to longer days and warmer weather.

Know >

1 What type of object is the Sun?

2 How long does it take for the Earth to complete one orbit of the Sun?

3 How long does it take for the Earth to complete one rotation on its axis?

Apply >>

4 Explain why the Sun appears to be higher in the sky during the summer months and always remains lower in the sky during the winter months.

Extend >>>

5 Mars has an axis of rotation which is 25° away from the vertical, but Venus has an axis of rotation which is 3° away from the vertical. What effect will this have on the day length and height of the midday Sun in the sky during a year on each of these planets?

Enquiry >>>>

6 Describe how you could use a table lamp and a globe model of the Earth to explain to a younger student how seasons are caused.

Key fact

→ The distances between the planets are *huge*.

» Core: The solar system

The scale of the solar system

A celestial body (or astronomical body) is any object in space. The largest celestial body in our solar system is the Sun, which is a star. It makes up 99.9% of the mass in the solar system. There are many other celestial bodies in the solar system, including eight planets, several dwarf planets, many moons, asteroids and other smaller objects.

Because the Sun is so massive, the gravitational attraction between the Sun and each planet is strong enough to hold the planets in orbit around the Sun. Gravity is also the force that keeps moons orbiting around their planet.

It is hard to visualise the distances and sizes involved in the solar system. The image on page 160 shows the relative sizes of the Sun and the eight planets, but the distance between them is not to scale. The image below shows the distances between the planets, but the size of the planets is hugely exaggerated.

▲ The distances between the planets in the solar system. The sizes of the Sun and planets have been enlarged to make them visible on this scale.

There really is a lot of space in space! For example, there is enough space between the Earth and the Moon to fit in all seven of the other planets without them touching, and still have space for a dwarf planet like Pluto! Between the four rocky inner planets (Mercury, Venus, Earth and Mars) and the gas giants (Jupiter and Saturn) is a huge area of space that contains the asteroid belt. The two outermost planets are the ice giants (Uranus and Neptune).

Orbits and rotation

All of the planets orbit the Sun in the same direction and in the same plane (two-dimensional region of space) which is called the **ecliptic**. This means that they are a bit like the rings around Saturn. The time taken for a planet to orbit the Sun once is known as its year length, which can be measured relative to Earth years.

The planets all rotate (spin) as they travel around the Sun. One rotation of a planet is the length of a day on that planet. The direction that they spin in, and whether their axis of rotation is at right angles to their orbit, varies from planet to planet. This is summarised in the table below.

Planet	Year length (in Earth years)	Day length (in Earth days)	Direction of rotation	Angle of axis of rotation from the vertical (degrees)
Mercury	0.24	176	Anti-clockwise	0
Venus	0.62	243	Clockwise	3
Earth	1	1	Anti-clockwise	23
Mars	1.9	1	Anti-clockwise	25
Jupiter	11.9	0.42	Anti-clockwise	3
Saturn	29.5	0.44	Anti-clockwise	27
Uranus	84	0.71	Clockwise	98
Neptune	165	0.67	Anti-clockwise	28

The further the planet is from the Sun, the longer it takes for it to orbit the Sun, so its year is longer. Mars rotates on its axis at the same speed as Earth, so its day is the same length. Mercury and Venus rotate very slowly. In fact, a day on Mercury is longer than a year on Mercury! The gas and ice giants rotate very quickly on their axes.

Planets which have an axis of rotation which is vertical relative to their orbit around the Sun have no noticeable seasons. Earth, Mars, Saturn and Neptune have similar seasonal variations. Uranus is very unusual, because it spins on its side, so its seasons are very strange indeed!

Worked example

Venus spins in a clockwise direction, whereas Earth spins in an anticlockwise direction. What difference would this have to an observer who was standing on Venus?

An observer on Venus would see the Sun rise in the West and set in the East, whereas on Earth the Sun rises in the East and sets in the West.

Know >

1 Which planet is closest to the Sun?

2 Which is the largest planet?

3 What force keeps the planets in orbit around the Sun?

Apply »

4 Explain why there is no seasonal variation of day length on Jupiter.

5 Which planet has the longest day?

Extend »»

6 Suggest why it takes longer for Uranus and Neptune to complete an orbit of the Sun than it does for the inner planets like Earth and Mars.

▲ The Moon can sometimes be seen from Earth during the day

» Core: Phases of the Moon

You will have noticed that the appearance of the Moon in the sky changes. Sometimes there is a full moon, sometimes there is a crescent moon and sometimes you cannot see the Moon at all. The Moon is quite often visible in the day time as well.

The lunar cycle

The images below show the way that the Moon's appearance changes during the course of a lunar cycle, which lasts approximately 28 days. Each row represents a seven day week.

▲ The phases of the Moon during a lunar cycle

From top left, the first few images are **crescent moons**. As the visible part of the Moon is growing, we say that the Moon is **waxing**. The image top right is a **half moon**, which is visible after seven days.

In the second row, you can see a **waxing gibbous** moon. A gibbous moon is when more than half is visible. The **full moon** is visible at the end of the second row, at 14 days.

In the third row (the third week of the lunar cycle), you can see a **waning gibbous moon**. Waning is when the size of the visible moon is decreasing. Another **half moon** is visible at 21 days.

In the fourth row, there is a **waning crescent moon**. On day 28, there is a **new moon**, when all of the side of the Moon which faces the Earth is in shadow, so it is very hard to see the Moon at all.

What causes these changes?

The way that the Moon orbits the Earth causes the changes to its appearance. Half of the Moon is always lit up by the Sun, but how much of this face of the Moon we can see from Earth depends on where the Moon is in its orbit around the Earth.

Common error !

At positions 4 and 8 in the diagram, the Moon, Earth and Sun do not line up completely, otherwise there would be either a solar eclipse (at position 8) or a lunar eclipse (at position 4).

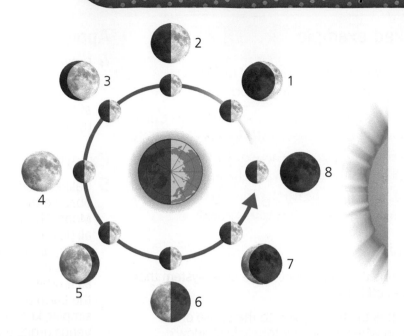

▲ A diagram showing how the orbit of the Moon around the Earth causes the lunar cycle (not to scale)

When the Moon is at position 1, an observer from Earth sees that most of the Moon is in shadow. This is a **waxing crescent.** When the Moon is at position 2, we see a **half moon.** At position 4, the side of the Moon that faces the Earth is fully lit, so we see a **full moon.** Position 5 is a **waning gibbous moon,** and position 7 is a **waning crescent.** At position 8, we see a **new moon,** as the side of the Moon facing us is completely in shadow. Remember that the Earth spins once every day, so everyone on Earth gets a chance to see the Moon at each of these positions. When looking at the diagram, don't worry too much about day and night!

How does the Earth appear when viewed from the Moon?

Astronauts have been able to take many photos of the Earth from space. Look at the image taken by astronauts on board Apollo 8 in 1968 as they orbited the Moon. Can you work out where the Sun must be?

The Sun is what is lighting up one side of the Earth, so the Sun must be above the Earth in this picture, and slightly to the right.

▲ Earth-rise as viewed from a spacecraft orbiting the Moon

Worked example

Explain why the appearance of the Moon changes during a 28 day period?

It takes 28 days for the Moon to orbit the Earth. During this time, the part of the Moon which is lit by the Sun may be facing the Earth (a full moon) or facing away from the Earth (a new moon), or it may be partly facing the Earth (crescent, half moon, gibbous moon).

Know >

1 What is the only object in the solar system that gives off light?

2 What is the name given to the phase of the Moon when it appears totally in shadow?

3 What name is given to the phase of the Moon when we can see only a thin edge of it?

Apply >>

4 Draw a diagram to show how observers on Earth can see a half moon. Start by putting the Sun at the centre of your diagram and the Earth at the top of your diagram. Finish your diagram by putting the Moon in the correct place.

Extend >>>

5 Galileo used one of the earliest telescopes to observe that Venus had phases, much like the Moon. This evidence ruled out the idea that all planets and the Sun orbited the Earth. Use a diagram to show how the appearance of Venus could change when viewed from Earth, as it orbits the Sun on a path which is between the Earth and the Sun. To make your diagram simpler, keep the Earth in a fixed place whilst Venus orbits the Sun.

>> Core: Beyond the solar system

Measuring distances in space

The distances in space are very large indeed, so our normal units of metres or kilometres become difficult to use. The Sun is 150,000,000 km away from Earth, and the second closest star is Proxima Centauri at a distance of 40,000,000,000,000 km.

Astronomical units (au) are often used to measure distance in space. This is equal to the distance between the Sun and the Earth. So Neptune is 30 au from the Sun. Proxima Centauri is 270,000 au from the Sun.

Another unit of distance which is used to measure very large distances is the **light year**. This is the distance that light can travel in one year. Because light travels very quickly, it only takes 8 minutes to reach us from the Sun. A light year is a very large unit of distance, equivalent to 9.5 million million kilometres. Proxima Centauri is 4.24 light years from the Sun.

> **Key word**
>
> A **light year** is the distance that light travels in one Earth year.

Unit	Useful for...
kilometre (km)	Measuring diameters of planets and stars, measuring distances between a planet and its moons
astronomical unit (au)	Measuring the distances within the solar system
light year (ly)	Measuring the distance between stars and galaxies

Galaxies and the universe

The Sun is one of an estimated 100–400 billion stars in the **galaxy** called the **Milky Way.** Many of these stars will be orbited by **exoplanets**. The Milky Way is approximately 100,000 light years in diameter, and it is shaped like a flattened spiral. Scientists do not know what it at the centre, but it seems likely that it is a **supermassive black hole.**

Scientists estimate that there are at least 2 trillion galaxies in the observable **universe**! It may take billions of years for light to reach Earth from other galaxies.

Exploring space

It is very expensive and time-consuming to explore space because of the huge distances involved. The furthest manned space flight was Apollo 13 which travelled around the back of the Moon. No humans have ever been to another planet within the solar system. We have sent robots to land on our nearest neighbour planet, Mars. We have sent unmanned space probes to observe up close every planet in the solar system and some probes have been sent outside the solar system. The first manned space explorations to Mars are currently being planned, scheduled to take place in the 2030s. Are you interested? It might be a one-way trip!

▲ The Milky Way

Worked example

What unit of distance would be most appropriate to measure the distance to Betelgeuse, a star in the constellation of Orion?

It would be best to measure the distance to Betelgeuse in light years.

Know >

1 What is the name of the nearest star to Earth which is outside our solar system?

2 What is the name of the galaxy where the Sun can be found?

3 How long does it take for light to reach Earth from the Sun?

Apply >>

4 Why do some scientists think that it would be better for the first manned mission to Mars to be a one-way trip?

5 Why is it preferable for most space research to be done by robots instead of humans?

Extend >>>

6 What do you think are the chances that life exists on exoplanets? Explain your answer.

▲ A self-portrait taken by the Mars Curiosity rover in 2012

» Extend: Making predictions at different latitudes

Latitude and longitude

Your location on Earth can be given in a number of ways. For example, if you are telling someone where you live, you might give them your address and postcode. If you were in a more remote part of the UK you could use an Ordnance Survey grid reference.

▲ A grid reference can provide an exact location within the UK

Elsewhere in the world, an exact location can be given using the latitude and longitude system. This system uses imaginary lines which separate up the surface of the Earth and can be used as a system of coordinates.

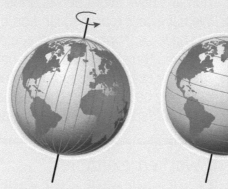

▲ Using latitude and longitude to describe a position on Earth. The diagram on the left shows lines of longitude. The diagram on the right shows lines of latitude

Key word

The **equator** is an imaginary line that runs around the Earth and separates the northern and southern hemispheres.

Longitude lines go from the north pole to the south pole, and can be used to show how far east or west you are from Greenwich, London, UK (which is taken to be 0° longitude). **Latitude** lines go in hoops around the Earth, with the largest one being at the **equator**, and this is called 0° latitude. 90° north is the north pole, and 90° south is the south pole. Buckingham Palace has a latitude of 51.5° north, and 0.1° west (because it is very close to Greenwich).

Shadows

During the daytime, if the Sun is visible in the sky, it will cast a shadow of an object on the ground. This principle has been used for centuries to tell the time using sundials, because as the Sun appears to move across the sky (due to the rotation of the Earth) the position of the shadow changes.

But the nature of the shadow will also change according to the season.

▲ The shadow created by the Sun can be used to tell the time on a sundial

Life near the equator

When you are on the equator, there is very little seasonal variation in day length or the Sun's intensity. The Sun appears to move at right angles to the horizon between sunrise and sunset.

Life near the poles

As the latitude rises and you move further from the equator towards either the north pole or the south pole, the seasonal variations become more extreme. Shadow lengths increase, seasonal variation in day length increases, and seasonal variation in the Sun's intensity increases. At positions north of the Arctic Circle or south of the Antarctic Circle, there is at least once per year 24 hours of continuous day time (during the summer) and 24 hours of continuous night time (during the winter). Near to the poles, shadows during this time will be very long, as the Sun will remain very low in the sky, and the intensity of the Sun will be very weak.

Tasks

1. Predict and justify how the length of the shadow on a sundial will change as a result of the season.
2. How will shadow length at the equator compare with shadow length at other latitudes?
3. What do you think life would be like north of the Arctic Circle or south of the Antarctic circle?
4. Make a prediction about what it might be like to celebrate Christmas at the British Antarctic Survey Halley Research Station, which has a latitude of 76° S and a longitude of 27° W.

▲ The BAS Halley Research Station in Antarctica

Enquiry:
Changing ideas about the solar system

Like so many areas of science, our understanding about the solar system has changed over time. Scientific understanding is always developing. Sometimes new observations or evidence comes to light which cannot be explained by an existing theory, so a new theory must be developed. At other times, a new theory is developed first, and then evidence is sought that will support or disprove it.

Early ideas about the solar system

Many ancient civilisations thought that the Earth was flat because they had no evidence to suggest that it was spherical. However, from the 6th century BCE (Before Common Era, or before the birth of Jesus), the idea of a spherical Earth was appearing in Greek philosophy.

From ancient times, people observed that the Sun rose in the sky in the morning and went down in the evening. The simplest explanation was that the Sun was orbiting the Earth, and this early model is called the **geocentric model** of the solar system. The Moon, planets, and all stars were also thought to be orbiting the Earth. Ptolemy helped to improve the geocentric model in the 2nd century CE (of the Common Era, after the birth of Jesus) with some complex adjustments that helped to explain the complicated movements of planets when viewed from the Earth.

▲ The Ptolemaic model of the solar system was geocentric

The heliocentric model

In 1543 CE, the Polish astronomer Copernicus published a detailed **heliocentric model** which was effective in explaining observations of celestial bodies and allowing accurate predictions to be made. In the heliocentric model, the Sun is placed at the centre of the solar system and all planets orbit the Sun in a fixed order. Copernicus was not the first person to suggest a heliocentric model, but his was a more detailed and accurate model, so he gets the credit for it. Copernicus explained that the seasons were caused by the tilt of the Earth's axis. However, Copernicus did not have any new evidence that proved his theory.

Key words

In the **geocentric model** of the solar system, all planets and stars orbit the Earth.

In the **heliocentric model** of the solar system, all planets orbit the sun.

Evidence for the heliocentric model

At the time, people did not want to believe the heliocentric model because it seemed to be less obvious than the geocentric model and also seemed to contradict the Bible. Scientists including Galileo used telescopes to discover that Jupiter had moons, which could not be explained by the geocentric model. The Sun was also seen to rotate on a fixed axis.

In 1687, Isaac Newton proposed the idea of gravity which explained what caused the motion of the planets around the Sun.

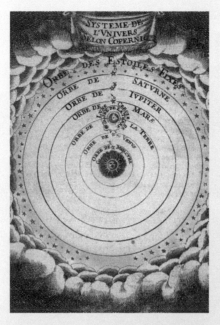

▲ The Copernican model of the solar system was heliocentric

▲ Galileo was one of several scientists to use telescopes to provide evidence which supported the heliocentric model

?

❶ Why was the geocentric model of the solar system the first one which was widely believed?

❷ Why did people not agree with the heliocentric model of the solar system at first?

❸ Copernicus devised his heliocentric model around 1510, but waited over 30 years before publishing it. It was printed the same year as he died. Suggest why he waited so long before publishing it.

❹ For the geocentric model, what came first: the theory or the evidence?

❺ For the heliocentric model, what came first: the theory or the evidence?

❻ Research how the work of Galileo and Kepler provided support for the heliocentric model. Write a letter from one of these scientists to someone who believes in the geocentric model to explain why they should change their beliefs.

Organisms

Learning objectives

15 Movement

In this chapter you will learn ...

Knowledge

- that the human skeleton supports and protects us, allows us to move and makes our blood cells
- that antagonistic muscle pairs work together by contracting and relaxing
- the definitions of the terms joints, bone marrow, ligaments, tendons, cartilage and antagonistic muscle pairs

Application

- how to explain how parts of the skeleton relate to the way it works
- how to explain why some organs contain muscle tissue
- how to explain how antagonistic muscles move joints

Extension

- how to describe the benefits and risks of improving human movement using technology

16 Cells

In this chapter you will learn ...

Knowledge

- that multi-cellular organisms are arranged into cells, tissues, organs and organ systems
- about the structures and features of specialised cells
- how to use a light microscope to observe and draw cells
- about the components of both plant and animal cells
- the definitions of the terms cell, uni-cellular, multi-cellular, tissue, organ, diffusion, structural adaptations, cell membrane, nucleus, vacuole, mitochondria, cell wall, chloroplast, cytoplasm, immune system, reproductive system, digestive system, circulatory system, respiratory system and muscular skeletal system

Application

- how to explain why multi-cellular organisms need organ systems to keep their cells alive
- how to use features of cells to suggest what kind of tissue or organism they come from
- how to use a microscope to identify and compare different types of cells
- how to explain how uni-cellular organisms are adapted to carry out functions that in multi-cellular organisms are done by different types of cell

Extension

- how to explain how medical treatments work based on cells, tissues, organs and organ systems
- how to suggest how damage to, or failure of, an organ would affect other body systems
- how to explain how the structure of different cells is related to their function

15 Movement

Your knowledge objectives

In this chapter you will learn:
- that the human skeleton supports and protects us, allows us to move and makes our blood cells
- that antagonistic muscle pairs work together by contracting and relaxing
- the definitions of the terms joints, bone marrow, ligaments, tendons, cartilage and antagonistic muscle pairs

See page 175 for the full learning objectives.

▲ Small battery-powered toys

▲ A mouse in a maze

▲ Sunflower flowers move during the day to always face the Sun

» Transition: What makes an organism alive?

The first two pictures below show some small battery-powered toys and a mouse in a maze. At first glance they might seem to be doing the same things. We know that the mouse is alive and the toys are not. But why?

There are seven life processes that separate a living organism from all dead or non-living things. Living organisms do all seven. Dead organisms no longer do these processes and non-living things never did and never will do them.

Movement

All living things move. It is often easy to see **movement** in animals as they hunt or avoid being hunted. But how do plants move? Well they do but often much more slowly than animals. For example, sunflowers rotate their flowers to face directly towards the Sun throughout the day.

Respiration

You will learn about **respiration** of animals, plants and microorganisms in Pupil's Book 2, Chapter 17. This crucial chemical process releases energy for all the other life processes from food.

Sensitivity

Animals, like humans, have many **senses** including sight, hearing, taste, touch and smell. These senses allow us to respond to

▲ How many sense organs can you see on a human face?

changes in our environment. Plants share senses with us. They are, for example, able to detect light and grow towards it. Our eyes allow us to detect light too. Plants can detect other factors. Roots grow downwards because of gravity.

Nutrition

Nutrition means feeding. Herbivores are animals that eat plants and so get their nutrition from them. Carnivores are animals that consume other animals. Omnivores, like many humans, eat both animals and plants. Plants themselves get their nutrition by making glucose in photosynthesis.

Excretion

All living organisms produce waste, which they need to remove from their bodies. Animal and plant cells produce carbon dioxide as a waste product of respiration. This is excreted from lungs when you breathe out. Other waste products are filtered from your blood by your kidneys and are excreted in your urine.

Reproduction

Reproduction means having offspring. Under normal circumstances, living things are able to reproduce. You will learn more about reproduction in humans and plants in Chapters 18 and 20.

Growth

All living things grow. This can range from the smallest single-celled bacterium to the largest tree or blue whale.

Many people remember the seven life processes by the first letters of their names which can be arranged (as in the chapter) to spell **MRS NERG**.

▲ Lions are carnivores that hunt prey for food

★ **You will learn about photosynthesis in Chapter 18 of Pupil's Book 2.**

★ **You will learn about respiration in Chapter 17 of Pupil's Book 2.**

Worked example

How do plants move?

Plant roots grow into the ground. Plant shoots grow up towards the light. Some plants, such as sunflowers, twist to follow the light.

Know >

1 What are the seven life processes?

2 What letters do many people use to remember the seven life processes?

3 HEXBUGS are small robotic animal toys. What processes do toys like HEXBUGS not complete?

» Core: Skeleton

The human skeleton is an amazing thing. You were born with around 300 bones in your skeleton. Many of these join together in childhood, particularly those in your skull. This reduces the number of bones in an adult to 206.

Bones themselves are organs like your heart and lungs. They come in a variety of shapes and sizes, to match their different functions. Many bones provide support and help us move, others protect our vital organs, and others still make our blood cells. Bones are lightweight yet incredibly strong. They contain calcium which gives them their strength. Calcium is an important mineral in the diet of children, who especially need it because their bones are still growing.

★ **You will learn more about diet in Pupil's Book 2, Chapter 16.**

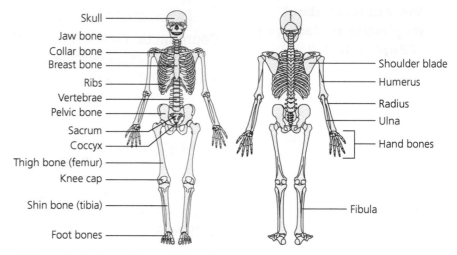

Skull
Jaw bone
Collar bone
Breast bone
Ribs
Vertebrae
Pelvic bone
Sacrum
Coccyx
Thigh bone (femur)
Knee cap
Shin bone (tibia)
Foot bones

Shoulder blade
Humerus
Radius
Ulna
Hand bones
Fibula

▲ The major bones in the human body

The major bones of the human skeleton are shown in the figure above. The largest bone in your body is the thigh bone or femur and the smallest is called the stapes and is found in your middle ear. Without this and several others you couldn't hear.

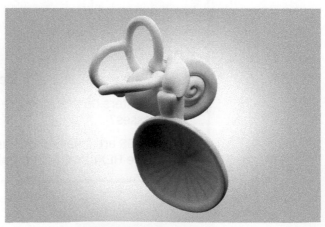

▲ The middle ear bones

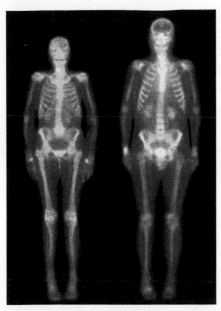

▲ Differences between male and female skeletons

There are differences in some bones and their position in the skeletons of men and women. In general the male skeleton is larger and more robust so men often have larger, stronger bones. As well as this, the shapes of male and female skulls are slightly different. The pelvis bones of women are wider to allow a baby to pass through during childbirth.

Support

Your skeleton provides a framework which **supports** your entire body. The bones in your feet and legs help you stand. The bones in your back and neck support your body and head. Without bones we would simply not be able to function as organisms. Your ribs surround the vital organs within your chest. Without ribs, your lungs would collapse and you would be unable to breathe and so die.

Movement

The calcium in your bones makes them incredibly strong. However you are not made from one giant inflexible bone. Your bones have many joints between them. These allow you **movement**. Without a joint in your shoulder and elbow you could not move your arm. Without joints in your wrists and between the bones in your thumb and fingers you could not move your hands. Joints in your skeleton allow you to move your limbs and your whole body from place to place.

Key word

Tendons connect muscles to bones.

Bones do not move by themselves however. Muscles are the only tissues in your body that can contract or relax to move. Your bones are therefore attached to some of your muscles by **tendons**. When these muscles contract or relax your skeleton moves.

Protection

▲ Dogs love eating the bone marrow from long bones

The strength of bone is well known. Dogs find it difficult to chew through them. It often takes a severe blow in sport or a fall to break a bone. The bones in your skeleton therefore provide you with **protection**. Your skull surrounds your brain. A break here can cause brain damage. The vertebrae bones that make up your backbone surround the nerves in your spinal cord. A break here can cause paralysis. Your rib cage surrounds your lungs, heart, and many other vital organs. All these need protecting from damage too.

Key word

Bone marrow is the tissue found inside some bones where new blood cells are made.

Blood cells

Why do dogs love chewing large bones? They enjoy eating the **bone marrow** that is in the middle of them. Bone marrow is a soft tissue that is found in the middle of some of your large bones. These are called **long bones** and include the bones of your arms and legs.

★ **You will learn more about respiration in Pupil's Book 2, Chapter 17.**

Your **red blood cells** and some **white blood cells** are made in the bone marrow of your long bones. In fact, they are so good at this that they can produce approximately 500 billion blood cells per day if needed. Your red blood cells carry oxygen around your body for respiration. Your white blood cells are part of your immune system and fight off infection by bacteria, fungi and viruses.

Worked example

How do the bones in your hand help you to move?

They are attached to muscles, which relax and contract to move your hand.

Know ›

1 How many bones are there in the human body?

2 What mineral helps strengthen bones?

3 Which is the largest bone in your body?

4 What bones protect your heart and lungs?

5 What types of cell are made in your bone marrow?

Apply »

6 What difficulties might a woman with small pelvic bones have?

Extend »»

7 What structures do other organisms have instead of bones?

8 How might cancer of the bone marrow affect a patient?

Enquire »»»

9 How can sports injuries affect your joints?

Key words

Joints are the points at which bones meet.

Cartilage is the smooth tissue found at the end of bones, which reduces friction between them.

- Pivot joint
- Ball and socket joint
- Hinge joint
- Ball and socket joint
- Hinge joint

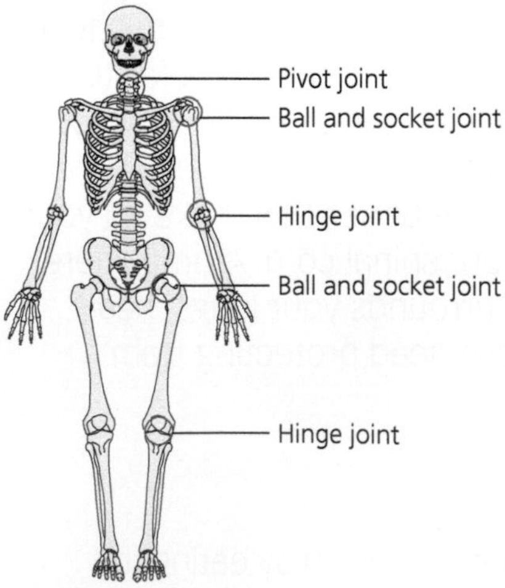

▲ The location of different types of synovial joint in the human body

» Core: Joints

Joints are parts of the human skeleton where bones meet. There are different types of joint in your body and they each allow different types of movement.

Synovial joints

The most common type of joint in your body is called a **synovial joint**. This type of joint also has the largest range of movement. There are different types of synovial joints, the locations of which are shown in the figure below.

All synovial joints have the same basic structure, as shown in the diagram on the next page.

The calcium in bones makes them very strong. However if bones in your joints rubbed against each other when you moved they would eventually wear away. So the bones in synovial joints don't actually touch each other. The ends of bones in synovial joints are covered with a strong and very smooth substance called **cartilage**. This is the same material that makes your nose feel hard.

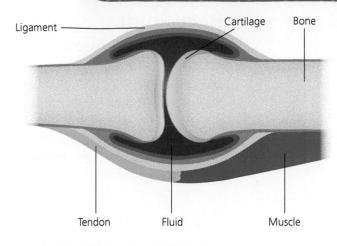

Ligament — Cartilage — Bone

Tendon — Fluid — Muscle

▲ A diagram of a typical synovial joint

A liquid called synovial fluid is found between the cartilage of the bones in synovial joints. This reduces any friction between the cartilage even further.

The bones in a synovial joint are held together by a tough, stringy tissue called **ligaments**. Sports men and women can tear these ligaments if they over-extend their joints. This form of injury is quite common amongst footballers stretching their bodies to shoot, block or tackle.

Eventually the cartilage does wear away in some older people. Then bones can actually rub against each other. This is a medical condition called arthritis and can be very painful. Joints can swell up and movement can become difficult.

Key word
Ligaments connect bones in joints.

Hinge joint

Hinge joints act like hinges attached to a door. They allow movement in just two directions. Your elbow is a good example of a hinge joint. This is the joint between your humerus bone in your upper arm and your ulna in your lower arm. Although it might feel like you can move your elbow in a circular motion, this comes from the movement of your shoulder and the bones in your lower arm. You can actually only move your elbow like a hinge in a door.

The knee is a modified hinge joint which allows some rotation as well as allowing your upper and lower leg bones to move like a hinge. This is called a **compound joint**. The knee is also the largest joint in your body.

Pivot joint

Pivot joints allow two bones to rotate about each other. There are two bones in your lower arm called the radius and ulna. You learned above that the ulna is joined to the humerus in a hinge joint to your upper arm. The ulna and radius bones in your lower arm are joined by a pivot joint. Without this joint you would not be able to keep your upper arm still, and rotate your lower arm, wrist and hand. You will learn more about this in the enquiry spread in this chapter.

Ball and socket joint

Your hip joint is an example of a **ball and socket joint**. Here the head, or end, of the femur bone in your leg has a ball shape. This fits into the cup-like socket of your pelvis bone. There is much greater rotation in ball and socket joints than either hinge or pivot joints. So you are able to move your leg in many more directions.

Fixed joints

Fixed joints are not synovial joints. They do not allow movement. You learned on page 178 that some of the bones in your skull fused together after you were born. This reduced the total number of bones in your body to 206 from approximately 300. The joints between the bones in your skull are fixed joints.

Damage to joints, bone of muscle

Regular exercise is extremely important for good health. However, sometimes injuries do occur. These are often because people have not 'warmed up' by doing gentle stretching exercises first. Bones can become cracked or broken. Joints can become dislocated when the end of a bone is no longer in its correct position. Tendons that hold muscles and bones together can become torn. Many footballers injure their achilles tendon, which connects the calf muscle to the heel bone. Muscles can be over-extended and damaged. Finally, ligaments, which connect bones on either side of joints, can become over-stretched. Damage to the cruciate ligaments, which surround the knee, are another common sporting injury.

These injuries are likely to cause pain, and often swelling and bruising. To treat injuries to joints, they should be rested, have ice applied to reduce the swelling, be compressed by a support bandage and elevated (lifted) to prevent swelling.

Worked example

How are elbows and knees different?

Elbows are hinge joints, which allow movement in two directions. Knees are compound joints, which are like hinge joints but allow some rotation.

Know >

1 What are joints?

2 What is the most common type of joint in your body?

3 Where in your body is a hinge joint?

4 Which is your largest joint?

5 What reduces friction in synovial joints?

Apply >>

6 How would the movement of your arm be different if your elbow was a ball and socket joint?

Extend >>>

7 Why does lifting a swollen joint reduce its size?

Enquiry >>>>

8 Are there any other types of joint found in different animals?

>> Core: Antagonistic muscle pairs

You learned earlier in this chapter that bones are unable to move by themselves. The only tissues in your body that can move are muscles. Therefore to allow you to move some of your muscles are connected to your bones by tendons.

However muscles are only able to **contract and relax**. When they contract they become shorter and fatter. This pulls on the tendons that are attached to bones and so therefore moves them. But there is a problem here. Muscles only have fine control of contraction, not relaxation. So, when muscles contract they can do this in a range of speeds but always in a controlled way. When muscles relax they cannot do this with any real control. So we need pairs of muscles called **antagonistic muscles** to work together.

> **Key word**
>
> An **antagonistic muscle** pair is when muscles are working in unison to create movement.

Antagonistic muscles in your arm

Imagine you are at the gym holding a dumbbell about to exercise the muscles in your upper arm. Your whole arm is by your side and you are about to move your lower arm at the hinge joint in your elbow. As you lift the dumbbell from down by your hip to your shoulder your biceps muscle contracts. You are able to do this in a slow and controlled manner.

You can also straighten your arm, and you can push against resistance when you do this. So there must be another muscle working here and it is called your triceps muscle. It is found in the back of your upper arm.

When your biceps muscle contracts as you pull the dumbbell upwards, your triceps muscle relaxes. When your triceps contracts, your biceps relaxes.

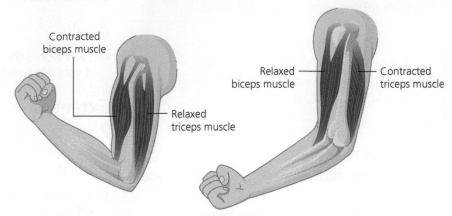

Contracted biceps muscle

Relaxed biceps muscle

Contracted triceps muscle

Relaxed triceps muscle

▲ Antagonistic muscle pairs in the upper arm

Antagonistic muscles in your leg

Now imagine you are doing kick ups with a football. As you move your lower leg upwards, your quadriceps muscle contracts. As before, you are able to do this in a slow and controlled manner. Your quadriceps muscles are the large muscles at the front of your thigh.

Now if you only had your quadriceps muscle your foot would only be able to fall to the floor. You would not be able to place it down carefully to regain your balance. But you can do this in a controlled manner, so there must be more muscles working here. They are called your hamstring muscles and are found at the back of your thigh.

When your quadriceps muscles contract as you lift your leg upwards, your hamstring muscles relax. When your hamstring muscles relax, your quadriceps muscles contract. This allows you to carefully move your leg up and down to keep doing kick ups with the football.

Forces exerted by different muscles

It is difficult to compare the force exerted by different muscles in your body because of their different shapes and sizes. Your jaw muscle is possibly the strongest. You need this force to break up tough food. Other muscles in your body are not as strong but have vital functions. Your gluteus maximus muscle is one of your hip muscles and is the largest in your body. It keeps you standing upright. Your heart muscle will beat over three billion times in your life. The muscles in your eyes move thousands of times per hour to allow you to see. The muscles of a women's uterus are strong enough to push out a baby.

In Pupil's Book 2 you will learn about turning forces. These rotate around a pivot. The size of the force is calculated by multiplying the force by the distance to the pivot. A person with a longer arm has a greater distance and so should be able to apply a greater turning force in an arm-wrestle.

Worked example

Why do muscles work in antagonistic pairs?

Muscles can only contract or relax. Working in antagonistic pairs allows controlled movement in both directions.

Know ❯

1 What are the two processes that muscles can do?

2 What name is given to muscles that work in pairs?

3 What are the names of the muscles that work antagonistically in your upper arm?

4 What are the names of the muscles that work antagonistically in your leg?

Apply ❯❯

5 What would happen without antagonistic muscle pairs?

Extend ❯❯❯

6 Can you make a model of a moving arm using everday materials like those below?

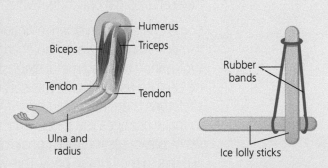

Humerus
Biceps
Triceps
Tendon
Tendon
Ulna and radius
Rubber bands
Ice lolly sticks

▲ Making a model of antagonistic muscle pairs in the upper arm

Enquiry ❯❯❯❯

7 How similar are muscles in other mammals?

» Extend: Technology for improving human movement

★ **You will learn more about classification of vertebrates in Chapter 17.**

For many years science-fiction film directors have written scripts for movies set in the future in space. To live in space on spaceships and travel the huge distances that the actors do in many of these films, the technology available in the future is obviously greater than our own. In some of these films people wear mechanical suits that sit over their clothes. These often make them faster and stronger.

▲ Mechanical suit in Aliens©

▲ Mechanical suit in Avatar©

▲ Mechanical suit in Ironman©

Internal and external skeletons

You know that your bones are inside your body. So you have an **endoskeleton**. All mammals (including us), reptiles, birds, amphibians and fish have endoskeletons. In fact all of these organisms have an internal backbone which is why they are called vertebrates.

Some characters in science-fiction movies including Wolverine© in the X-men© franchise have had changes made to their endoskeletons. Wolverine© has had metal extensions to the bones in his hands which he can use as weapons. This sort of medical enhancement to human skeletons is entirely invented. We do not currently have the technology to make changes like this. Even if we did, scientists and politicians are likely to express massive concern about the ability to make changes like this to the human body. This would become an ethical issue that some people disagreed with for religious or moral reasons.

Some organisms don't have internal bones forming an endoskeleton. They have a protective structure on their outside called an **exoskeleton**. Because these organisms do not have a backbone they are called invertebrates. Insects, crustaceans (like crabs) and molluscs (like snails) are common examples of this type of organism. They have an exoskeleton structure that surrounds their entire body. This provides support and protection like your endoskeleton. (The tortoise has both an endoskeleton and an exoskeleton shell to hide in.)

▲ Exoskeletons of insects, crabs and snails

You learned earlier in the chapter that bones are organs made stronger by calcium. Insect exoskeletons are made from a very strong polymer called chitin. This substance is a little like that which makes our finger and toe nails.

Many of the exoskeleton suits seen in movies like Aliens© and Avatar© don't seem to cover the entire human body for protection. Ironman© is a notable exception. Our police and armed forces already have Kevlar© body armour, or bullet-proof jackets, to do this. The exoskeleton suits seem to act like external bones and make the wearer faster and stronger. They must be powered to do this of course.

The future

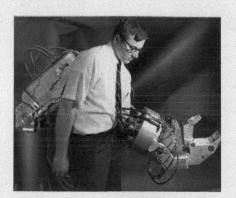

How long will it take scientists to make a human exoskeleton like we see in the movies? Well the technology already exists. In 1960 the US Army made a suit called Hardiman, which made lifting over 100 kg feel like lifting less than 5 kg. However, the suit has a mass of over 600 kg, which means that it is very difficult to move when wearing it. More recent work by the US Army is developing a 'future soldier' suit that will do the same thing but at a fraction of the mass. How long will it be before soldiers are wearing exoskeletons?

▲ Exoskeletons of Hardiman in the 1960s and the US 'future soldier'

But the uses of this sort of exoskeleton suit are not just limited to fighting. They would be able to replace limbs from people born without them or those that have had to have amputation operations. Exoskeleton suits could help many people in this or similar situations to move again. They could make movement easier for older people with reduced mobility. These suits could also help the people of the emergency services like firefighters who work in dangerous conditions.

Tasks

1. **Where** are exoskeletons found?
2. Name three types of organism with exoskeletons.
3. What element makes animal bones strong?
4. What are insect exoskeletons made from?
5. How could an exoskeleton suit help a firefighter?
6. Who else could benefit from wearing an exoskeleton?

Enquiry:
How similar are chicken wings and human arms?

There are some very obvious differences between chicken wings and human arms. Chicken wings are much smaller and covered with feathers for example. But remove the feathers and adjust the sizes to be similar, and then how different are they?

▲ Externally a chicken wing and human arm appear very different

Bones of the human arm and chicken wing

Your arm extends from your shoulder to your wrist. The bones in your arm are shown here. The upper part of your arm is made from one bone called the humerus. This joins your shoulder in a ball and socket joint which can rotate in many more directions than other joints. The top of the humerus is shaped like a ball and it fits into the cup-like structure of the scapula or shoulder bone.

Below your humerus is your elbow which is an example of a hinge joint. Below this, and extending to your wrist are two bones called the radius and ulna. The ulna is fixed in position, but the radius can rotate around the ulna. This joint is called a pivot joint and allows more movement than the hinge joint. It is this joint that allows you to turn your arm to, for example, unlock a door.

Interestingly, a similar pattern is seen in the bone structure of a chicken wing. You will learn how similarities in the pentadactyl limb in vertebrates is evidence for evolution in Pupil Book 2. Here we will not focus on the digits found in toes and fingers but the main bones of the arm.

Like our upper arm the upper wing of the chicken has a single bone called the humerus. Similarly, the shoulder joint of a chicken joins the shoulder to the humerus and allows the wing to rotate. As in humans, this is a ball and socket joint.

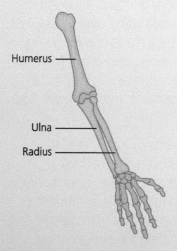

Humerus

Ulna

Radius

▲ Bones in a human arm

Below the humerus is the chicken wing elbow. Like ours, this is a hinge joint and joins the humerus to the two bones below. These are called the radius and ulna. As in humans, the ulna is fixed in position, but the radius can rotate around the ulna. This is a pivot joint, and allows the chicken wing more movement than a hinge joint.

★ **You will learn more about classification in Chapter 17.**

There are striking similarities in the bone structure of a human arm and chicken wing. Both organisms belong in the vertebrate group of the animal kingdom, so perhaps this similarity is not so surprising.

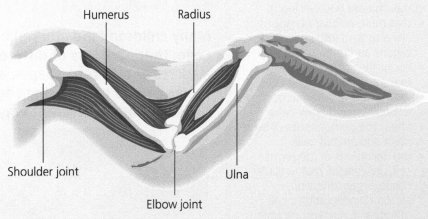

▲ Bones in chicken wing

Muscles

You learned earlier in this chapter that muscles can only contract and relax, so need to work in **antagonistic pairs** for controlled movement. You learned that the movement of the lower arm in humans is controlled by the triceps and biceps muscles.

We see exactly the same pattern of antagonistic muscles in the upper chicken wing. The biceps muscle is attached to the ulna in the lower part of a chicken wing, so when it contracts it moves this lower part upwards. While this is happening, the triceps muscle relaxes. To move the lower part of the wing down again, the biceps muscle relaxes and the triceps muscle contracts. This is attached to the very top of the radius and ulna and pulls them in line with the humerus. This lowers the bottom of the wing.

❶ What is the name of the bone in your upper arm?
❷ What are the names of the two bones in the lower part of a chicken wing?
❸ What type of joint is your shoulder?
❹ What type of joint is found in the middle (elbow) of a chicken wing?
❺ What does the pentadactyl limb provide evidence for?

16 Cells

Key word

A **cell** is the simplest unit of a living organism and it contains parts which carry out the seven life processes.

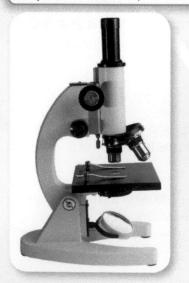
▲ A microscope can be used to see cells

» Transition: The building blocks of living things

Many children (and adults!) enjoy building things out of Lego. There are lots of different Lego bricks. They come in all sorts of different shapes, sizes and colours. All sorts of things can be made using Lego, by combining the different bricks in lots of different ways.

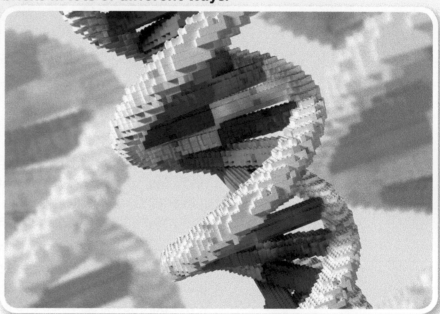
▲ Lego bricks can be used to build a model of DNA

All living things like plants and animals are made from tiny building blocks called **cells**. There are many different types of cell, and they can be arranged in many different ways. This creates billions of different species of living things. Remember that living things carry out the seven **life processes** (MRS NERG): Movement, Reproduction, Sensitivity, Nutrition, Excretion, Respiration and Growth.

Many living things, like plants and animals, are made from lots of cells, but some living things are made from only one cell. Bacteria are living things which are made from just one cell each.

How big are cells?

Cells are almost always too small to see with the naked eye. Biologists use **microscopes** to see cells. Microscopes are instruments that magnify tiny objects, much more powerfully than a magnifying glass.

What is inside a cell?

Cells are made of different parts which have different jobs to do. Not all cells contain the same parts, but these parts are present in almost all animal and plant cells.

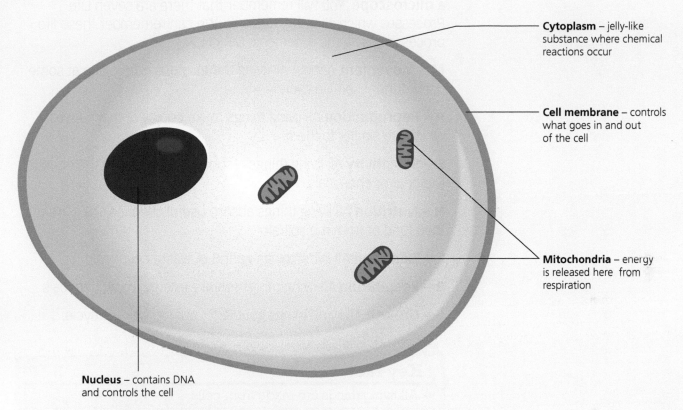

Cytoplasm – jelly-like substance where chemical reactions occur

Cell membrane – controls what goes in and out of the cell

Mitochondria – energy is released here from respiration

Nucleus – contains DNA and controls the cell

▲ The structure of a typical animal cell and what each part does

Common error

You will learn in chemistry that atoms contain a nucleus in the middle. The nucleus of a cell is much much bigger. Atoms are much smaller than cells.

Worked example

Almost all animal cells contain a nucleus. Describe what is so important about the nucleus and explain why almost all cells need one.

The nucleus contains the DNA and controls the cell. Without a nucleus, most cells will not be able to do their job properly.

Apply >

1 All cells have a cell membrane around the outside of the cytoplasm. Describe what the membrane does and why it is so important.

2 Which is smallest and which is largest: an apple, a cell or the nucleus of a cell?

3 Why would you feel tired with fewer mitochondria in your cells?

» Core: Comparing cells

Cells are the building blocks of living things. All living things are made from cells, which are too small to see unless you use a **microscope**. You will remember that there are seven Life Processes which all living things do. We can remember these life processes using the mnemonic MRS NERG…

M – Movement Almost all living things move from place at some point during their life cycle.

R – Reproduction All living things make copies of themselves which are similar to them.

S – Sensitivity All living things respond to changes to their habitat, like heat or chemicals.

N – Nutrition All living things absorb useful chemicals, e.g. from their food or from the soil/air.

E – Excretion All living things get rid of waste chemicals.

R – Respiration All living things release energy from chemicals.

G – Growth All living things get bigger during their life cycle.

> **Key fact**
> → All living things are made from cells.

A typical animal cell

There are lots of different types of animal cell, but most of them have the following parts. You can see a simple diagram of an animal cell on the previous page.

Nucleus – contains the genetic material (DNA). DNA contains instructions which control the cell and allow it to copy itself.

Cytoplasm – a jelly-like substance which makes up most of the cell and where chemical reactions occur.

Cell membrane – surrounds the cytoplasm and controls the movement of substances in and out of the cell, often by **diffusion**.

Mitochondria – this is where respiration takes place, releasing energy from chemicals obtained from food.

Key word

Diffusion is when one substance spreads out from a high concentration, through a gas or a liquid, due to the random movement of particles.

A typical plant cell

A typical plant cell looks like this and contains all the same parts as an animal cell, plus a few more.

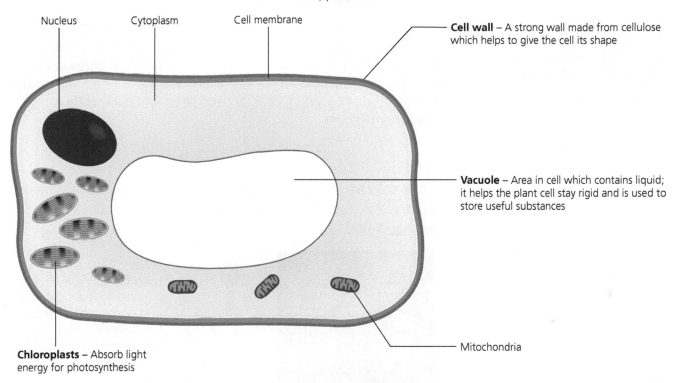

Nucleus   Cytoplasm   Cell membrane

Cell wall – A strong wall made from cellulose which helps to give the cell its shape

Vacuole – Area in cell which contains liquid; it helps the plant cell stay rigid and is used to store useful substances

Mitochondria

Chloroplasts – Absorb light energy for photosynthesis

▲ The structure of a typical plant cell and what some of the parts do

Worked example

Why is it important for all living things to respond to changes in their environment?

Living organisms must respond to changes in their environment to avoid being damaged by harmful things, or to benefit from useful things like food or energy sources.

Know >

1 List the parts of a cell that are present in both plant and animal cells.

2 Which part of the cell controls what enters and leaves the cell?

Apply >>

3 Why is it important for all living things to excrete their waste products?

4 When animals eat, they obtain nutrients from their food which can be used in their cells in respiration to release useful energy. How do plants get these useful chemicals which are a store of energy?

Extend >>>

5 Suggest which part of a plant cell from a leaf gives it its green colour.

6 Muscle cells contain a lot of mitochondria. Explain why.

Enquiry >>>>

7 Name the piece of apparatus which is used to study cells in school laboratories.

8 Investigate how electron microscopes work. There are two types: TEM and SEM.

Key word

Structural adaptations are features of a cell which allow it to do its job.

» Core: Specialised animal cells

The cells we have studied so far show the typical parts of a cell, and clearly show the difference between plant and animal cells. But within a plant or an animal, there are lots of different types of cell, and they have specific **structural adaptations** to help them to do their job. The photographs below have been artificially coloured to make them clearer.

Sperm cell

Sperm cells are the male reproductive cell in all animals. They need to carry DNA from the male to meet the female reproductive cell (the egg/ovum).

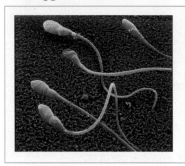 

Long tail to allow the cell to swim to meet the egg. Lots of mitochondria at the base of the tail to release energy through respiration.

Egg cell

The egg cell (ovum) is the female reproductive cell in all animals and also contains DNA. In mammals, it remains inside the female's body during fertilisation.

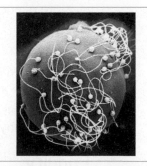 

Large cytoplasm contains lots of stored chemical energy to allow for the embryo to begin growing.

Mammalian red blood cell

Red blood cells carry oxygen around the body, using a red pigment called haemoglobin.

No nucleus, to make more space for lots of haemoglobin in the cytoplasm, to carry lots of oxygen.

Nerve cell (neurone)

Nerve cells carry electrical signals from one part of our nervous system to another. They communicate with other nerve cells using chemical signals.

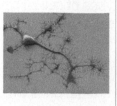 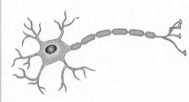

Branching structure allows it to communicate with other nerve cells to pass messages.

Ciliated epithelium cell

These cells line certain tubes within the body and move tiny particles from one place to another by waving their tiny hair-like fingers. They are found in the airways, oviducts and uterus.

 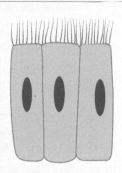

Hair-like fingers waft dust particles out of lungs in mucus, and waft the egg through the female reproductive system.

Worked example

Describe the structure of a ciliated epithelium cell and explain how this helps it to function effectively in the lining of the oviduct.

The side of the cell which faces into the uterus is covered in tiny hair-like structures. These move forwards and backwards in waves. This helps the lining of the uterus move the egg from the ovary into the uterus.

Know >

1 Which kind of cell has a branching structure to allow it to communicate with other cells?

2 Which kind of cell has a tail to help it to swim?

Apply »

3 Describe the structure of a sperm cell and explain how this helps it to function effectively.

Extend »»

4 Why is the structure of red blood cells in mammals unique?

» Core: Specialised plant cells

Plants contain lots of specialised cells as well as animals. Again, the different types of cell all have structures or features which help them to do a specific job.

Palisade leaf cell

Palisade cells are the ones which do most of the photosynthesis.

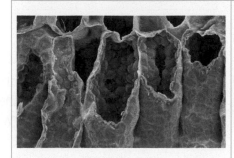

Contain chloroplasts to absorb lots of light and produce glucose in photosynthesis.

Root hair cell

Root hair cells are on the roots and absorb water and minerals from the soil.

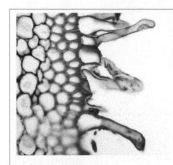
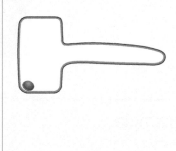

Have a large surface area to absorb water. No chloroplasts are needed.

The pink cells on the right are root hair cells

Phloem cell

Phloem cells join up end to end to make tubes in which sugar solution can move around in the plant.

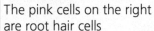
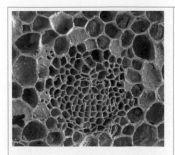

The ends of the cells have holes in them to allow sugars to move from one cell to the next.

The green cells are phloem cells

Xylem cell

Xylem cells are dead cells and have no ends. They join up end to end to make continuous tubes which allow water to move quickly through the plant.

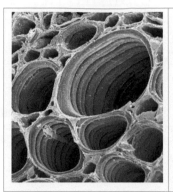 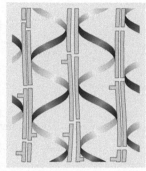

The cells are strengthened with a substance called lignin.

Worked example

Describe the structure of a root hair cell and explain how it is adapted to its function.

A root hair cell has a very large surface area. This allows it to absorb water and minerals effectively, which is important in the roots of a plant.

Know >

1 What is the function of xylem in a plant?

2 What is transported in the phloem system within a plant?

Apply >>

3 Describe the structure of a palisade leaf cell and explain how it is adapted to its function.

Extend >>>

4 Why do root hair cells not contain any chloroplasts?

>> Core: Organising cells

Uni-cellular organisms

Some simple organisms, like bacteria, are made from just one cell. We call these organisms **uni-cellular**. There is a huge variety of single celled organisms, so we will focus here on a typical bacterium. Individual bacteria cells must be able to do all of the life processes independently, without other cells to help. The diagram shows some of the adaptations they might have. One of the biggest difference between a bacterium cell and a cell from an animal or plant is that it has no nucleus. However, some other uni-cellular organisms do have a nucleus, like yeast for example.

> **Key word**
> A **uni-cellular** organism is a living thing made from just one cell.

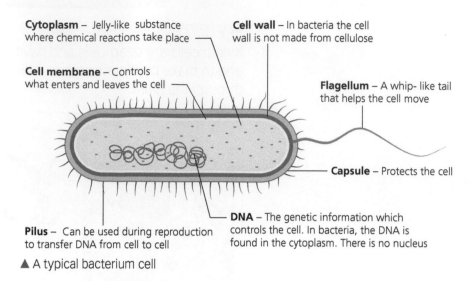

Cytoplasm – Jelly-like substance where chemical reactions take place

Cell wall – In bacteria the cell wall is not made from cellulose

Cell membrane – Controls what enters and leaves the cell

Flagellum – A whip- like tail that helps the cell move

Capsule – Protects the cell

Pilus – Can be used during reproduction to transfer DNA from cell to cell

DNA – The genetic information which controls the cell. In bacteria, the DNA is found in the cytoplasm. There is no nucleus

▲ A typical bacterium cell

Multi-cellular organisms

Many living things, including all plants and animals, are made from many cells which are joined together. We describe these kinds of living things as **multi-cellular** organisms. The cells work together to complete the life processes. A group of specialised cells in the same place which are similar to each other is called a **tissue**. Muscle tissue is made from lots of muscle cells. Fat tissue is made from lots of fat cells. Nervous tissue is made from lots of nerve cells.

Different tissues which are in the same place and work together make an **organ**. The organ has a specific job to do, and the tissues work together to make that happen. The heart is an organ which pumps blood around the body. It is made from muscle tissue and nervous tissue. The leaf is an organ in plants which specialises in photosynthesis. The flower is the reproductive organ in flowering plants.

Different organs work together in a **system,** to perform one or more life processes. Multi-cellular organisms need their systems to stay alive. Their cells, tissues and organs would not be able to survive on their own. Uni-cellular organisms are self-sufficient.

Here are some of the systems in the human body, and their main organs.

Key words

A **multi-cellular** organism is made from many cells. All animals and plants are multi-cellular organisms.

A **tissue** is a group of cells of one type.

An **organ** is a group of different tissues which work together to carry out a job.

System	Function	Organs
Circulatory system	Transports substances around the body	Heart, blood vessels
Digestive system	Breaks down and then absorbs food molecules	Oesophagus, stomach, intestines, liver, pancreas, rectum
Reproductive system	Produces sperm and eggs, and supports a growing foetus	Female: ovaries, oviducts, uterus, vagina Male: testes, sperm tube, penis
Immune system	Protects the body against infections	Skin, bones, glands (e.g. tonsils)
Respiratory system	Replaces oxygen and removes carbon dioxide from the blood	Lungs, trachea (windpipe), diaphragm
Muscular-skeletal system	Supports, moves and protects the body	Bones, muscles

Worked example

Describe the function of the cell membrane and explain why cells from both uni-cellular and multi-cellular organisms need this part of the cell.

The cell membrane controls which substances enter and leave the cell. It also holds the cytoplasm in. Without it, cells would not be able to keep useful chemicals inside, and harmful chemicals outside the cell.

Know >

1 Which system in the human body obtains nutrients from food?

2 Name the organs in the circulatory system.

3 What is an organ?

Apply >>

4 List the similarities between a bacterium cell and a plant cell.

5 List the differences between a bacterium cell and a plant cell.

6 Is this cell from an animal or a plant? Suggest the tissue or organ that it came from. Explain your answer.

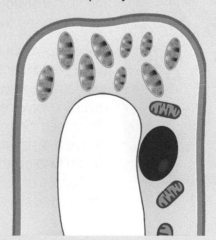

Extend >>>

7 The sperm cell is the only animal cell that has a flagellum. Explain why this is necessary, and why no other animal cells need one.

Enquiry >>>>

8 See if you can find out how bacteria use their pili (the plural for pilus) to share sections of DNA during reproduction.

» Extend: Healthcare and the human body

As our understanding of cells, tissues, organs and systems has developed through scientific research, so has our ability to treat injuries and illnesses.

Early medical beliefs and treatments

The ancient Greeks and Romans believed that there were four liquids in the human body which they called the four **humours**. The four humours were **blood, phlegm** (mucus), **black bile** and **yellow bile.** Bile is a digestive juice produced by the liver. People used to believe that each person had to have the right balance of these four humours in their body, or they would be ill.

Discovering bacteria

In the mid-1800s, French scientists including Louis Pasteur described bacteria and suggested that they caused diseases in animals, including humans. This discovery prompted improvements in hygiene such as washing hands before surgery and childbirth, and improved survival rates. But it was the accidental discovery of the first **antibiotic,** penicillin, by Alexander Fleming in 1928, which revolutionised the treatment of bacterial diseases.

Antibiotics have led to huge improvements in disease treatment since World War 2 because they kill bacteria (or stop them multiplying). However, bacteria reproduce so quickly that they quickly evolve **resistance** to antibiotics, so new ones have to be developed. Scientists cannot keep up with how fast bacteria can evolve a resistance to antibiotics, and many experts at the World Health Organization believe that this is one of the greatest scientific challenges of the modern era. Diseases which we thought we had under control may prove deadly in years to come.

▲ Alexander Fleming, 1881–1955

In-vitro fertilisation

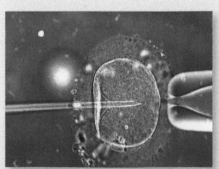

Some human couples find it hard to get pregnant through normal sexual intercourse. Understanding how cells work has allowed scientists to make fertilisation happen outside of the body, which is **called *in-vitro* fertilisation,** or **IVF.** An egg cell is taken from the woman's ovary and mixed with sperm outside the body, typically in a Petri dish, to allow fertilisation to take place. If the man's sperm are not good at swimming, a single sperm may be injected into the cytoplasm of the egg using a very fine needle, viewed under a microscope.

▲ In one form of IVF treatment, a sperm is injected into an egg cell so that fertilisation occurs

The fertilised egg is incubated for a few days to make sure it is developing correctly, and then injected back into the mother's uterus. The first baby conceived using IVF was born in 1978, and since then more than 5 million children have been born using this technology across the world.

Bone marrow transplants

Bone marrow is a tissue found inside bones. In the largest bones in the body, it is responsible for making red blood cells and some white blood cells, producing around 2 million blood cells every second! One form of **cancer** is called **leukaemia,** where the production of abnormal white blood cells gets out of control.

One treatment for leukaemia is to destroy the patient's bone marrow using **chemotherapy** or **radiotherapy** so that it can no longer make blood cells. Then, healthy bone marrow from a donor is **transfused** into the bloodstream of the patient. Bone marrow cells move from the blood and colonise the interior of the bones, and start producing new red and white blood cells.

Recreational drugs

Our knowledge of cells, tissues, organs and systems has developed to allow us to identify the effects of taking drugs. Of course, not all drugs are harmful. Many drugs are medicines, which help improve health. These are either bought from the shops or prescribed by doctors. Recreational drugs are those used without a medical reason. Drugs can be divided into four categories, shown in the table below.

Type of drug	Effects	Legal examples	Illegal examples
Stimulants	Increases nerve activity so speeds up reaction times	Nicotine, caffeine	Ecstasy, cocaine
Depressants	Reduces nerve activity so slows reaction times	Alcohol	Solvents, cannabis
Hallucinogens	Alters what a person sees and hears	None	LSD
Painkillers	Reduces pain	Aspirin, paracetamol	Morphine (without prescription)

Tasks

1. Find out online what MRSA stands for, and why health authorities are so concerned about it.
2. How do you think scientists disproved the theory of the four humours?
3. Why is it important to wash your hands, or use hand sanitising gel when you enter hospital wards? And when you leave?
4. Find out which organs and tissues can be transplanted successfully.
5. In 2016, Professor Sergio Canavero announced that he was planning to perform the first 'head transplant' before the end of 2017. See what you can find out online about this. Try to explain why it would be more accurate to describe the procedure as a 'body transplant'.

Enquiry:
Using a light microscope

You can use a light microscope in school to observe your own cells. We use the term **light** microscope because this kind of microscope makes an image by shining light through a tissue sample, not because it is easy to lift it up!

The easiest and safest cells to observe from your own body are cheek cells, which line the inside of your mouth.

Safety precautions

- Methylene blue solution is harmful if swallowed, and it may irritate your skin

- Microscope slides and cover slips are made from glass and may cut your skin, so handle them carefully and report any breakages

- Follow your teacher's instructions about how to dispose of the slides at the end of the lesson

Preparing the slide

1 Swab the inside of your cheek using a cotton bud. Swipe it from side to side for 10 seconds.

2 Smear the cotton bud onto the centre of a microscope slide for 2–3 seconds.

3 Your teacher will add one drop of methylene blue solution, which will stain the DNA in the nucleus of the cell.

4 Gently lower a cover slip on top of the drop of liquid so that there are no bubbles trapped underneath. Your teacher will show you how to do this.

5 Any excess liquid can be carefully blotted away using a paper towel but be careful not to touch the methylene blue solution.

▲ Collecting cheek cells

Observing your cells

6 Always start with the lowest magnification lens, which is often a 4× or 10× lens. (Remember, if you want to calculate the total magnification, you multiply the eyepiece lens magnification by the objective lens magnification, both of which are labelled in the diagram on page 203.).

7 Place the slide onto the stage (platform) of the microscope and secure it with the spring clips so that the sample of cells is underneath the lens.

8 Adjust the mirror or turn on the lamp so that light is passing through the sample.

9 Use the coarse focus knob to move the stage as close to the lens as possible. You should do this whilst looking at the gap between the lens and the slide from the side.

10 Look through the eyepiece and turn the coarse focus knob slowly to move the stage further away from the lens until the cells are in focus.

11 Use the fine focus knob to adjust the focus until you can clearly see the cells in the sample.

12 Draw a diagram of a few cells using a sharp pencil to show how they are arranged. Label your diagram with the date, what the sample is (cheek cells) and the magnification you are using.

13 Rotate the part of the microscope which contains the lenses so that you are using a higher magnification. Adjust the focus if necessary.

14 Draw a close up diagram of one cell. You should be able to see the nucleus which will be stained dark blue due to the DNA it contains. If you see small spots of dark blue in your sample, these might be bacteria, which contain their DNA loose in the cytoplasm. Bacteria cells are much smaller than human cells.

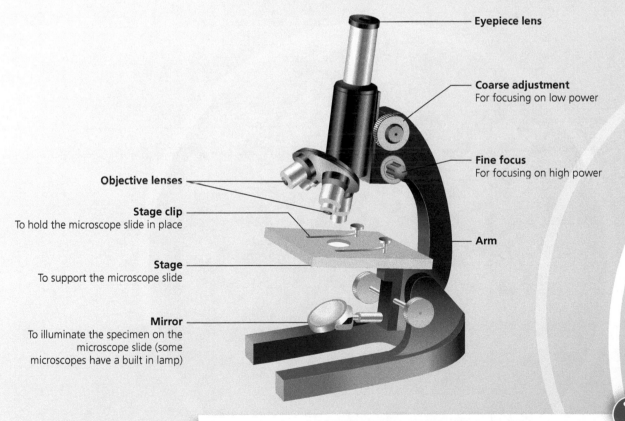

Eyepiece lens

Coarse adjustment
For focusing on low power

Fine focus
For focusing on high power

Arm

Objective lenses

Stage clip
To hold the microscope slide in place

Stage
To support the microscope slide

Mirror
To illuminate the specimen on the microscope slide (some microscopes have a built in lamp)

❶ How do we calculate the total magnification of a light microscope?
❷ What is the function of the mirror?
❸ What safety issues are there when using a light microscope?
❹ What part of an animal cell does methylene blue stain?
❺ When would you use the coarse focus and when would you use the fine focus?

Ecosystems

Learning objectives

17 Interdependence

In this chapter you will learn...

Knowledge

- that organisms in a food web depend on each other, so a change in one population leads to changes in others
- about the factors that can change the numbers of a species
- the definitions of the terms food web, food chain, ecosystem, environment, population, producer, consumer and decomposer

Application

- how to describe how the number of predators and prey changes over time
- how to explain the effects of environmental changes and toxic materials on the numbers of a species
- how to combine food chains to form a food web
- how to explain issues with human food supplies in terms of insect pollinators

Extension

- how to suggest what might happen if an unfamiliar species were introduced into a food web
- how to explain how toxic substances can accumulate in human food

18 Plant reproduction

In this chapter you will learn...

Knowledge

- how plants are adapted to disperse seeds using wind, water or animals
- that plants reproduce sexually to produce seeds, which are formed following fertilisation in the ovary
- that a plant's reproductive organs are found in the flowers
- that pollen can be carried by the wind, pollinating insects or other animals
- the definitions of the terms pollen, ovules, pollination, fertilisation, seed, fruit and carpel

Application

- how to describe the main steps in plant reproduction
- how to describe parts of the flower and explain their function
- how to suggest how a plant disperses its seeds based on the features of its fruit or seeds
- how to explain why seed dispersal is important to survival of the parent plant and its offspring

Extension

- how to describe similarities and differences between the structures of wind pollinated and insect pollinated plants

17 Interdependence

Your knowledge objectives:

In this chapter you will learn:

- that organisms in a food web depend on each other, so a change in one population leads to changes in others
- about the factors that can change the numbers of a species
- the definitions of the terms food web, food chain, ecosystem, environment, population, producer, consumer and decomposer

See page 205 for the full learning objectives.

» Transition: Classification

Classification is the process of putting things into groups and then giving them names. This is a process that we naturally find quite easy. When you were younger, do you remember putting your dolls into order of their height, or grouping your toy cars according to their colour? This is a simple form of classification.

There are around 8.7 million different types of life (or species) on Earth at the moment. We have not yet discovered a large proportion of these; perhaps over 90% of them. Wouldn't you like to be the person who discovered a species that is new to science?

But how do we know when we have found a new species? How do we know it isn't just a different looking version of an existing one? In fact, how do scientists estimate there at around 8.7 million species? Well, the first step is classification. We need to be able to identify individuals from any potential new species that might have been discovered. So we need a way of grouping organisms and then naming them. We cannot conserve any species from extinction if we do not first identify them.

Carl Linnaeus (1707–1778) was a Swedish scientist who is known as the pioneer of classification. He developed the idea of grouping organisms that are similar within groups, that are themselves in larger groups. Lions are a species of big cat. They are found in a larger group called panthera with tigers, jaguars and leopards. The group panthera then belongs to a larger group and so on.

◄ Carl Linnaeus was the pioneer of classification

The five biggest groups are called the kingdoms. These are: animals, plants, fungi, bacteria and single-celled organisms (which are a little like animals and a little like plants).

▶ Examples of the five major groups of life on Earth

Within the animal kingdom are two major groups. These are animals with a backbone (called vertebrates) and animals without (called invertebrates). Within the vertebrates are five groups. These are: amphibians, mammals, reptiles, birds and fish.

We humans are mammals and so are found within this group.

▲ Examples of the five major groups of vertebrates

Know ❯

1 What is classification?

2 Who was a pioneer of classification?

3 What are the five major groups of life on Earth?

4 What are invertebrates?

» Core: Feeding relationships

Food chains

Food chains are flow diagrams that show feeding relationships between **populations** of organisms in an **ecosystem**. An example of a woodland **food chain** is:

Oak tree → Squirrels → Foxes

Each stage of a food chain is called a **trophic level**. At the base of almost every food chain is a green plant or alga that completes a chemical reaction called **photosynthesis**. Organisms at the bottom of all food chains are called **producers**. They don't actually produce energy, but do convert the Sun's energy into a sugar called glucose in photosynthesis.

The first organism above the producer in a food chain is called a primary **consumer**. All primary consumers are therefore found in the second trophic level. They depend upon the producers for the nutrients they need to survive. They are animals and almost all are **herbivores** (plant or algae eaters).

Primary consumers are eaten in turn by secondary consumers, which are found in the third trophic level. Because these eat primary consumers, they are always **carnivores** (meat eating animals) or **omnivores** (animals that eat meat and plants or algae). Secondary consumers depend upon primary consumers for the nutrients they need to survive.

Consumers in higher trophic levels are called tertiary (third) and then quaternary (fourth). At the top of every food chain is a **top predator**. This organism is not preyed upon by any other.

The number of organisms in a population depends upon:

- the numbers of predators or prey it has
- the amount of disease and volume of pollution
- the competition for limited resources such as water and nutrients.

Food webs

It is quite unusual for only one food chain to be present in any location. Often several food chains fit together to make a **food web**.

A second woodland food chain is:

Oak tree → Caterpillar → Shrew → Owl

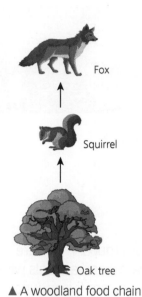

Fox

Squirrel

Oak tree

▲ A woodland food chain (not to scale)

★ **You will learn more about photosynthesis in Pupil's Book 2, Chapter 18.**

Key words

A **food chain** is part of a food web, starting with a producer, ending with a top predator.

A **population** is a group of the same species living in an area.

An **ecosystem** is the living things in a given area and their non-living environment.

A **trophic level** is a stage in a food chain or web.

Photosynthesis is the process that uses the Sun's energy to convert carbon dioxide and water into glucose and oxygen.

A **producer** is a green plant or algae that makes its own food using sunlight.

A **consumer** is an animal that eats other animals or plants.

A **food web** shows how food chains in an ecosystem are linked.

These two food chains can fit together to make a woodland food web:

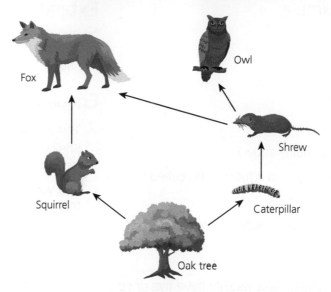

▲ The two food chains make a woodland food web (not to scale)

▲ This dung beetle is a decomposer

Decomposers

Up to this point we have only thought about living organisms and their part in food chains and webs. Some organisms die of natural causes or have parts that are not eaten by consumers. Organisms called **decomposers** help breakdown waste and dead animals and plants. They recycle their nutrients into the soil or water in their ecosystem. The most common examples of decomposers are fungi, but bacteria, dung beetles and earthworms are also examples.

Energy flow

The arrows in food chains and webs show the transfer of energy between the trophic levels. They do not show which organisms eat each other. The squirrels get the energy they need to complete the seven life processes from oak trees. The foxes get the energy they need from the squirrels.

Only about ten percent of the energy in each trophic level is passed to the one above. The organisms at the lower trophic level use the remaining 90% of energy to live. Because the energy transferred between trophic levels is only about 10%, most food webs only have around five or fewer trophic levels.

Worked example

Why are decomposers important?

They recycle nutrients and return them to the food chain.

Know >

1 What do food chains show?

2 What are the stages in a food chain called?

3 What do decomposers do?

4 Where are producers found in a food chain?

Apply >>

5 Why do food chains only usually have five or so trophic levels?

Extend >>>

6 Where on Earth do we find the only food chains that don't begin with a photosynthetic plant or alga?

Enquiry >>>>

7 This is a woodland food chain: Oak tree → Squirrels → Foxes. Describe what happens if a farmer kills too many squirrels.

8 Can you find a food chain with more than five trophic levels?

>> Core: Pyramids of number and biomass

Pyramids of number

The number of organisms at each trophic level in a food chain can be shown in a pyramid of number. An example for a marine food chain is:

Plankton → Sardines → Tuna → Sharks

A pyramid of number for this food chain looks like this.

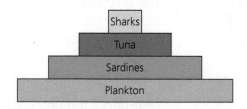

▲ A pyramid of number for a marine food chain

The bars must always be drawn with square ends and always be touching. The size of the bars represents the number of organisms at that trophic level. In this case there are more plankton than sardines, more sardines than tuna, and so on.

Pyramids of number are very often this shape. If there were more sharks than tuna, it is likely they would eat them all. There are two occasions when pyramids of number are not perfect pyramids. The first occurs when the producer is one large organism and not lots of smaller ones. In the rainforest food chain below, the mango tree is so large it can support all the toucans.

Mango tree → Toucan → Jaguar

So the pyramid of number looks like this.

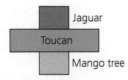

▲ A pyramid of number for a rainforest food chain with a small bar for the first trophic level

The second occasion when pyramids of number are not perfect pyramids occurs when the top predator is infected with parasites like fleas. These are usually very small organisms that can infest their host in large numbers. If the jaguar in the above food chain had fleas, the pyramid of number would look like this.

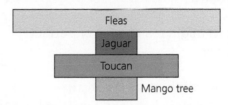

▲ A pyramid of numbers for a rainforest food chain with a large bar for the top trophic level representing fleas

Pyramids of biomass

Pyramids of number show very clearly the number of organisms in a food chain. However, they do not give any indication of the size of the organisms. Pyramids of biomass take this into account. **Biomass** is the dry mass of a living or recently dead organism.

This is the last food chain we looked at:

Mango tree → Toucans → Jaguar → Fleas

The pyramid of biomass for this food chain looks like this.

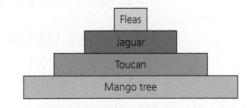

▲ A pyramid of biomass for a rainforest food chain

You can see that this pyramid of biomass has a large bar at the bottom because the mango tree is much larger than the combined biomass of the toucans. You can also see this pyramid has a tiny bar at the top because the total biomass of the fleas is less than the jaguar.

Because biomass can vary with the amount of water an organism consumes, scientists often measure biomass after it has been dried. Unfortunately, this means organisms have to be killed before they are dried out.

You have just read that a large organism at the bottom of a food chain or many smaller ones at the top can stop pyramids of number from being perfect pyramids. Pyramids of biomass are always perfect pyramids. If they were not, the ecosystem is in severe danger and is not likely to remain healthy.

Worked example

What two situations might result in an imperfect pyramid of number?

A large organism, such as a tree, at the bottom will give a small bar, and parasites, such as fleas, on the top carnivore will give a larger bar.

Know >

1 What do the arrows in food chains show?

2 What shape must the bars in a pyramid of number be?

3 What is biomass?

Apply >>

4 Under normal circumstances, which is the largest bar in a pyramid of number?

5 What is different about the shapes of pyramids of biomass and number?

6 Why do we dry out biomass samples before measuring them?

Enquiry >>>>

7 Research food chains in different parts of the world. For each draw a pyramid of number and biomass. Are they the same or different?

>> Core: Changes in ecosystems

Changes to food chains

Changes in any ecosystem can have significant effects on the organisms that live there. Think about this Artic food chain.

Plankton → Cod → Ringed seal → Polar bear

If the number of plankton was reduced, then the cod, seals and bears would have less to eat and could all die. If a large number of the cod were killed, then the seals and the bears would not have as much to eat and so could die. But the plankton would not be eaten by the cod, and so might increase in number. Finally, if the bears were killed, then the seals would not be being eaten so could increase in number. This could mean more cod are killed, which in turn might result in the increase in the amount of plankton.

Changes to food webs

You learned on page 208 that food chains are often joined together into food webs. So the following food chain can be added to the one above.

Plankton → Cod → Harbour seal → Killer whale

The food web would then look like this.

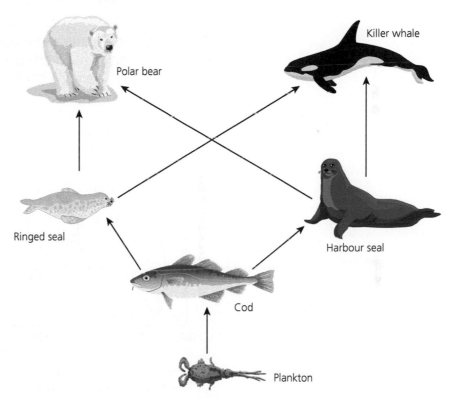

▲ The two food chains make an Artic food web (not to scale)

Changes in food webs are often more complicated and difficult to predict. Think about the food web above. If the number of ringed seals reduced, then the number of bears and killer whales might reduce. But they could eat more harbour seals, which would mean they reduced and the number of ringed seals stayed relatively constant.

Interdependence

We know that small changes in ecosystems can have huge consequences because all the living organisms depend upon each other in different ways. This dependence upon each other is called **interdependence**.

Healthy ecosystems are usually good at coping with small changes to the numbers of living organisms within them. However, and very sadly, the impact that we humans have on some ecosystems is too large for this to easily be undone.

Predator–prey cycling

Predators hunt and kill other animals for food. Prey are the animals killed by predators. Have you ever wondered why don't the predators kill all the prey? They would probably die if they did but this isn't the reason. Predators don't stop and think about the effects of killing prey. Their instincts take over.

The lynx and hare

The Canadian lynx is a small wild cat that lives in the mountains of Canada. It feeds on the snowshoe hare, which is similar to a large rabbit. For nearly a hundred years the numbers of lynx and hare were recorded. A famous graph of their numbers was produced.

★ **You will learn about the adaptations of predators and prey in Pupil's Book 2, Chapter 19.**

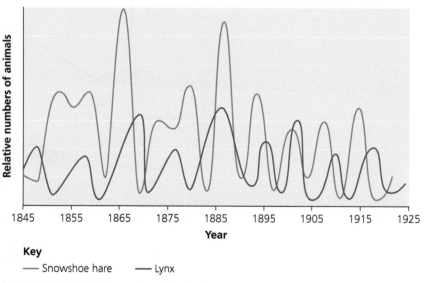

Key

—— Snowshoe hare   —— Lynx

▲ The predator prey cycling of the lynx and hare

This graph shows very clearly the predator–prey cycling we see in many different ecosystems. If we look at the beginning of the graph, we see the number of predators (lynx) increasing. Because of this, and at the same time, we see the number of prey (hare) decreasing. This is because there are more lynx eating more hare. You can see that around 1845 on the graph.

However, the number of lynx can only increase so far. There will come a point where the numbers of hare are not large enough to support all the lynx. There are simply not enough hare to eat. So the number of lynx will eventually start to fall. You can see that around 1850 in the graph.

As the number of lynx starts to fall the number of hare will increase. This is because there are fewer lynx to eat the hare. You can see this around 1850 on the graph.

Eventually the number of hare will have increased to support more lynx. So the lynx population increases again. You can see this around 1855 on the graph. We are now back at the beginning of the predator–prey cycle.

Throughout the cycle the number of predators usually remains less than the number of prey.

Worked example

What would happen to the producers and secondary consumers in a food chain if the number of primary consumers decreased?

The number of producers would be likely to increase and the number of secondary consumers would be likely to decrease..

Know >

1 What does interdependence mean?

2 What do predator–prey cycles show?

Apply >>

3 Why do the numbers of predators eventually fall?

4 Why do the numbers of prey eventually rise?

Extend >>>

5 What factors could affect the numbers of predators in a population?

6 Why do we usually find more prey than predators?

Enquiry >>>>

7 Can you find any food chains with only two trophic levels?

» Extend: Human impact on ecosystems

Pollution

Pollution is the release of substances into the **environment** which have a harmful effect on the organisms living there. Pollution can damage the surrounding air, water and land.

A major example of air pollution is carbon dioxide from the burning of fossil fuels, which is contributing to the greenhouse effect and causing global warming. Methane gas is another greenhouse gas. Hydrocarbons are released from burning fossil fuels and cause smogs. Sulfur dioxide is also released from burning fossil fuels and causes acid rain. Examples of water pollution include human waste as sewage, and nitrates and phosphates found in fertilisers. We are also causing land pollution by burying our waste in landfill. Huge holes are filled with our rubbish. How much of this could have been recycled?

★ **In Pupil's Book 2, Chapter 19 you will learn about how the air pollution from the industrial revolution caused a significant change in the populations of peppered moth.**

Bioaccumulation

Some particularly nasty polluting chemicals cannot be excreted from organisms. They bind closely to the tissues inside them and are not removed in urine or faeces. These chemicals are found in very low, and often unharmful, concentrations in organisms at lower trophic levels. However, because these chemicals cannot be excreted they build up in the organisms at higher trophic levels. We call this **bioaccumulation**. Sadly, these chemicals often have devastating effects at these higher concentrations.

A famous example of bioaccumulation was first seen in an insecticide called DDT. After the Second World War, the use of this pesticide by farmers increased significantly. As described above, this concentrated in organisms in the higher trophic levels. Predatory birds including the bald eagle and peregrine falcon were particularly badly affected.

Farming

Farming is an extremely important agricultural industry, for it provides foods and other materials that we require. Farmers are under pressure to keep their costs down and to maximise the amount of food they produce (their yield). To do this many farmers have turned to **intensive farming**. This is a range of

▲ A combine harvester is an example of machinery used in intensive farming

techniques which significantly increases yield but can damage local ecosystems. Intensive farming includes:

- the use of machinery
- keeping animals in small enclosures
- using pesticides and fertilisers
- removal of hedgerows to make tending to crops easier.

Of particular concern to many scientists at the moment is the large reduction we are seeing in the number of pollinating insects. In particular, the number of bees has reduced significantly. Intensive farming is partly responsible for this. Without these pollinating insects the security of our food is reduced. Food security exists when all people always have access to basic foods to meet their needs for an active and healthy life. In recent years, floods near factories that make biscuits have reduced their availability and poor weather in Southern Europe has reduced the availability of winter salad and vegetables. Food security is reduced by droughts, economic instability and armed conflict.

Not all farmers use intensive methods. **Organic** farmers do not use pesticides or chemical fertilisers. They rotate their crops between fields to avoid having to use fertilisers. They keep their livestock in a free range environment. As a result, organic food is often more expensive, but many people are willing to pay for it to help protect our environment.

▲ Princes Charles and William on their organic Duchy Home Farm

Key fact

➜ Insects are needed to pollinate food crops.

Tasks

1. What is bioaccumulation?
2. What is the famous example of bioaccumulation that resulted in weakened egg shells of predatory birds?
3. Why do some chemical pollutants bioaccumulate?
4. Why are organic foods more expensive than those produced by intensive farming?
5. Why are scientists so concerned about the reduction we see in pollinating insects like bees?
6. What would happen if the world ran out of fertilisers and pesticides?

Enquiry:
Introducing an alien species into an ecosystem

Invasive species are those that have been introduced by humans to new ecosystems. Because they have not evolved to live in these ecosystems, their introduction can often have devastating effects on the native animals and plants.

Red and grey squirrels

We currently have two species of squirrel in the British Isles. The first and original native species is the red squirrel which used to be found all over the country. In the 1870s, rich land owners chose to import the North American grey squirrel to their land. They did this because they thought it was fashionable. They did not predict the consequences of their actions. They grey squirrel is now an invasive species.

The grey squirrel is much larger than the red squirrel which means it is stronger and can survive harsher winters on its fat reserves. It has quickly spread from the estates and has outcompeted the red squirrel, which is now only found in small populations in North Wales, Northern England, Scotland and Ireland. The grey squirrel also brought a new virus which has killed many red squirrels.

▲ The size difference between red and grey squirrels

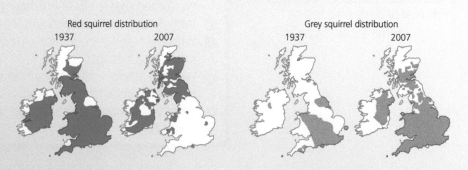

Red squirrel distribution
1937   2007

Grey squirrel distribution
1937   2007

▲ The spread of the red and grey squirrels between 1937 and 2007

The cane toad

The cane toad is a large land toad which is native to Central and South America. It was introduced into islands in the Pacific Ocean including Australia to reduce the numbers of cane beetles which were threatening crops. The cane toad reproduced quickly and has spread.

It possesses poison glands on its back which are highly toxic to organisms in Australia that try to eat it. These organisms, like the northern quoll and pet dogs, have not evolved with the cane toad and so are more easily killed by its poison. In Central and South America, where the cane toad originates, it has a number of natural predators like the caiman, and some snakes, eels and fish. These are not found in Australia.

The cane toad is larger than native Australian toads and is not killed by native predators. This means it can outcompete other toads, which are now threatened with reduced numbers.

▲ Numbers of the Northern quoll have reduced significantly in Australia since the introduction of the cane toad

| 1940 | 1960 | 1975 | 1980 |

▲ The spread of the Cane toad between 1940 and 1980

There is good news here though. Scientists have observed predators in Australia changing their behaviour to avoid the poison. Some crows now flip the toad over to kill and eat it, and the dwarf crocodile has been observed only eating the hind legs, so as to avoid the poison on its back.

Rhododendron

Plants can be invasive species as well as animals. Rhododendrons are very colourful flowering plants first imported from countries around the Mediterranean to the British Isles in the 1760s. Although many people do like the look of these plants, they have begun to spread and cause damage to the ecosystem in which they live. They reduce the number of earthworms in the surrounding soil. They also smother the growth of smaller native plants, which can reduce the numbers of birds and small mammals such as dormice.

▲ We see fewer herbivores such as dormice in areas in which rhododendrons have been planted or spread

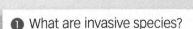

1. What are invasive species?
2. State an example of an invasive species.
3. Why do invasive species often cause so much damage?
4. Why is the grey squirrel outcompeting the red squirrel?
5. What other invasive species can you find out about?

18 Plant reproduction

Your knowledge objectives:

In this chapter you will learn:

- how plants are adapted to disperse seeds using wind, water or animals
- that plants reproduce sexually to produce seeds, which are formed following fertilisation in the ovary
- that a plant's reproductive organs are found in the flowers
- that pollen can be carried by the wind, pollinating insects or other animals
- the definitions of the terms pollen, ovules, pollination, fertilisation, seed, fruit and carpel

See page 205 for the full learning objectives.

★ **You will learn more about photosynthesis in Pupil's Book 2, Chapter 18.**

» Transition: Life cycle of flowering plants

Plants can be split into two groups: those that flower and those that don't. You will probably know lots of plants that do flower like daisies, daffodils and dandelions. But how many plants can you name that don't flower? Ferns and mosses are examples of plants that don't. But what about trees and grasses? Surprisingly to some people, both trees and grasses are examples of flowering plants. Their flowers are often relatively small and not colourful, especially in the case of grasses.

From seeds to seeds

Many plants begin their lives as a seed. When there is enough oxygen and water the seed will germinate. This means that it produces its first shoot which grows upwards towards the light and its first root which grows downwards following gravity. The germinated seed then grows into a small plant. The shoot continues to chase the light by growing upwards. It will start to produce leaves which are essential for photosynthesis. This process provides energy for the plant to grow. More roots will grow and existing roots will branch. They grow into the soil to hold the plant in its position and absorb more water for photosynthesis.

▲ The variation of flowers is huge

The plant will eventually reach reproductive maturity. This means the plant can now have its own offspring. To do this it will produce flowers. Flowers produce pollen which is the male sex cell of plants. It is their equivalent of sperm. Flowers also produce eggs. Pollination occurs when the pollen from one flower meets the eggs from the same or a different flower.

Some plants have evolved to spread pollen by insects. These plants have brightly coloured flowers, and can produce nectar (which is a sugary liquid) to attract pollinating insects. These move from flower to flower, drinking the nectar. As they move they transfer pollen.

Other plants like grasses are pollinated by the wind. They have less colourful flowers and don't produce nectar. They do not need to attract insects. Their pollen is simply blown by the wind from one flower to another.

Once pollination occurs the fertilised egg grows into a seed. The ovary that surrounded the egg grows into a fruit. Plants spread their seeds in a variety of ways. Some drop them in water. Others let the wind blow them to a new location. Others are caught on animals as they walk past them. Finally, the fruit surrounding some seeds attract animals which consume them. The seeds are them deposited in a new location when the animal produces droppings.

Worked example

How do flowers attract insects?

Flowers are often brilliantly coloured to attract insects. They produce nectar, which is a sugary liquid. Insects are attracted to flowers to collect the nectar.

Know >

1 What are the two types of plant?
2 What type of plant are grasses?
3 What are the male and female reproductive cells of plants?
4 What is the reproductive organ of plants?

Key words

Pollen contains the plant male sex cells found on the stamens.

Ovules are the female sex cells in plants found in the ovary.

Fertilisation is the joining of a nucleus from a male and female sex cell.

Pollination is the transfer of pollen from the male part of the flower to the female part of the flower on the same or another plant.

The **carpel** is the female part of the flower, made up of the stigma (where the pollen lands), style and ovary.

» Core: Sexual reproduction

Sexual reproduction involves the fusion of a male gamete or sex cell of a plant which is a **pollen** grain with the female gamete which is an **ovule**. This is **fertilisation** but is specifically called **pollination** in plants. You will learn how pollen can be spread by the wind or pollinating insects later in this chapter.

Flowers

There is massive variation in the colour and shape of flowers. These contain the reproductive organs of plants. Many flowers attract insects for pollination. Even plants like grasses have flowers. Here they are pollinated by the wind so the stamens and stigmas hang outside the flower. All flowers have a single purpose which is to allow pollination to occur.

The structure of a typical flower is shown in the diagram below. The functions of these parts are given in the table below.

▲ The structure of insect- and wind-pollinated flowers

Flower part	Function
Petals	Often brightly coloured to attract insects for pollination
Stamen	The male part of the flower made from an anther which produces pollen (the male sex cell) and the filament which holds the anther and the pollen in position
Carpel	The female part of the flower, made up of the stigma where the pollen lands, style and ovary
Stigma	Sticky part of a flower that collects pollen grains for pollination
Style	The style is a long thin stalk that connects the stigma and ovary
Ovary	Produces the female sex cell (egg) and grows into the fruit after pollination
Ovule	Contains the female sex cell (egg), which grows into the seed after pollination

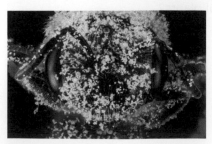

▲ The head of this bee is covered with pollen

Pollination occurs when a pollen grain (the male gamete) lands on the stigma. This can either be pollen blown by the wind in the case of grasses and other wind pollinated flowers or on a pollinating insect such as a bee. Insect-pollinated flowers are often brightly coloured, with an attractive scent, and usually contain sugary nectar to attract insects. They usually possess less pollen than wind-pollinated plants, but this is often sticky or spiky to attach to the insect. Plants with brighter, more scented flowers with more nectar are more likely to be pollinated. Wind-pollinated flowers do not need to attract insects so are often small, with plain colours and no scent or nectar. Pollen is produced in much greater amounts because more is lost. The pollen itself is very light and smooth to be blown in the wind. Plants that produce flowers with more pollen that is smoother and lighter are more likely to reproduce.

The pollen grain must be of the same species for pollination to occur. The pollen grain forms a pollen tube which grows down through the style towards the ovary. The nucleus of the pollen grain containing DNA passes down the tube to fertilise the egg inside the ovule. The fertilised egg develops into an embryo. The ovule grows into the **seed**. The ovary forms the **fruit** that surrounds the seed.

Seeds contain an embryo which possesses the DNA of a new plant. The seed itself provides nourishment for the embryo to grow into a new plant. Seeds can be stored for long periods of time before they germinate and begin to grow into a new plant. This ensures they only do so when the conditions are correct. Fruits surround the seed and provide protection. They also provide an incentive for animals to eat seeds and spread them around in their droppings.

Advantages and disadvantages

Sexual reproduction is extremely important in both animals and plants, but it is a time-consuming process which uses a large amount of energy. Animals need to find and attract a mate. Plants need to grow a flower to attract pollinating insects or catch pollen if wind pollinated.

So why do animals and plants expend this extra energy? You have already learned that sexual reproduction produces offspring that are not genetically identical to their parents or each other. This means that populations of animals or plants that that have been produced by sexual reproduction often have vastly more variation within them. This is called a wide **gene pool**. If a disease were to infect the organism or the population be threatened by a new herbivore (for plants), predator (for animals) or environmental change, then a population with a wide gene pool is far more likely to adapt and survive. Variation is a key principle in the process of evolution by natural selection. You will learn about this in Chapter 19.

> **Key words**
>
> A **seed** is the structure that contains the embryo of a new plant.
>
> A **fruit** is the structure that the ovary becomes after fertilisation, which contains seeds.

> **Key facts**
>
> → Flowers contain the plant's reproductive organs.
> → Pollen can be carried by the wind, pollinating insects or other animals.

Know >

1 What are the male and female sex cells in plants?
2 What is pollination?
3 What part of a plant attracts insects?
4 What is the function of the stamen?
5 What is the function of the stigma?

Apply >>

6 Why are scientists worried about the decline in the number of bees in the UK?

Extend >>>

7 Why is it good for plants to have their seeds dispersed far from them?

Enquiry >>>>

8 Find out about plants with the smallest and the largest flowers.

» Core: Asexual reproduction

Plants are able to reproduce asexually as well as sexually. You learned about the sexual reproduction of plants earlier in this chapter. Asexual reproduction only involves one parent and the offspring are clones of each other and the parent, so genetically identical.

Imagine if one of your fingers was cut off and this was able to grow into a clone of you. Some plants are able to do this. (Amazingly some animals like the starfish are also able to.) This is called vegetative reproduction if it happens in plants.

▲ A plant leaf producing clones by asexual reproduction

Some plants make small clones of themselves on their leaves. These grow into small plantlets and fall off. They can put down roots and grow into a new plant. Other plants like the spider plants and strawberries produce **runners**. These are thin, flexible stem-like structures that grow horizontally along the ground. Small plantlets develop on offshoots. This process has an added advantage of encouraging the clones to put down roots further from the parent plant. In this way the clones are less likely to compete for resources such as water, light and space. Plants that can produce more clones in asexual reproduction are more likely to have offspring.

Why don't all clones look the same?

Cloned plants, whether natural or artificial, may not look like their parent plant even though they have identical DNA. You will learn about the differences between genetic and environmental variation and why this is the case in Chapter 19.

Advantages and disadvantages

Asexual reproduction is a much quicker and often easier process than sexual reproduction, which requires a second parent and ability to join gametes or sex cells. It often requires less energy expenditure by the parent. Some species of animal like the water flea are able to reproduce asexually when environmental conditions are good, and sexually if they get worse.

▲ Strawberries producing clones on runners

If organisms only reproduce asexually then the population often has much less variation within it. The **gene pool** is narrower. If a disease were to infect the population, or if it were be threatened by a new herbivore (for plants), predator (for animals) or environmental change, then a population with a narrow gene pool is far less likely to adapt and survive. Variation is a key principle in the process of evolution by natural selection. You will learn about this in Chapter 19.

Worked example

Why do strawberry plants produce long runners?

So that the baby plantlets that grow on the runners settle away from the parent plant and don't compete with it.

Know >

1 What is asexual reproduction?

2 What are runners?

3 Why can asexual reproduction be an advantage?

4 Why can asexual reproduction be a disadvantage?

Apply >>

5 Why are some vets worried about inbreeding in some pedigree dogs?

Extend >>>

6 Why don't all cloned plants look the same?

Enquiry >>>>

7 Find out about other animals that can reproduce sexually and asexually.

>> Core: Seed dispersal

It is important the plants don't just drop their seeds beneath them. If they did and the seeds germinated and grew, they would be competing for the same water and sunlight that the parent plant needs. So plants have developed a range of different ways of spreading their seeds. This is called dispersal. Plants that can produce more seeds and more effectively disperse them are more likely to have offspring.

Wind dispersal

Some plants like the dandelion have evolved very light seeds that are blown by the wind. The seeds have a small 'head' which consists of a small stalk that spreads out into thin branches at its top. These get caught in even the lightest breeze and can be blown far away.

▲ Dandelion seeds being blown in the wind

Other seeds like those of the maple tree form 'helicopters'. These seeds are heavier than those of the dandelion and so don't usually travel as far. As they fall from the tree, the shape of their helicopter blades makes them rotate. They catch the wind as they do this and so often fall further from the tree than those without helicopter wings. Acorns from oak trees and conkers from horse chestnut trees don't have helicopters and so usually fall vertically, landing closer to the parent tree.

▲ Maple seeds and their flight path as they fall

Water dispersal

Other plants, such as coconut palms, produce seeds that float. These plants often grow near water. Coconut palms grow on tropical beaches all over the world and so drop their coconuts into the water. These can travel across oceans and germinate on a different continent from which they originated.

▲ Coconuts can float across entire oceans before germinating

The water lily is another example of a plant that produces seeds that are dispersed by water. The water lily roots itself in a shallow pond or river and has leaves and flowers that float. The leaves get more sunlight for photosynthesis and the flowers can be pollinated by insects. When the seeds are formed they detach from the parent plant and float downstream.

Animal dispersal

Many plants produce a fruit full of natural plant sugars to surround their seed or seeds. This is to attract an animal to eat it. The seeds have protective coatings to mean they are not broken down in the animal's digestive system. They pass through their stomach and intestines and are released in their droppings. The time it takes for this to happen ensures the seed is more likely to be dispersed far from its parent.

Other plants produce seeds with hooks on. These are called burs. They catch onto the fur or hair or any passing animal. They are often removed hours later when the animal cleans itself or its fur naturally falls out. Burdock is a common plant which produces seeds with burs.

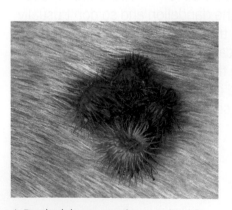

▲ Burdock burrs caught on animal fur

George de Mestral (1907 to 1990) was a Swiss scientist who invented the hook and loop fastener called Velcro. He returned from a walk with his dog and removed the burdock seeds that had fastened themselves to his clothing and his dog. He looked at them under the microscope and saw the burs. From this natural inspiration, he then invented Velcro.

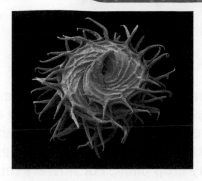
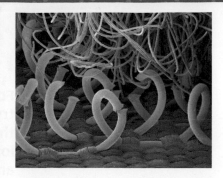

▲ The similarity between burdock burs and Velcro

Dispersal by ejection

Some plants like germaniums and euphorbias have developed a more explosive way of dispersing their seeds. They eject them forcibly from their pods. The side of the seedpod that faces the Sun dries out more quickly than the shaded side. This causes the pod to curl up and buckle. Eventually the seed pods snap back into shape and force out the seeds. This can make a sound loud enough for us to hear.

Worked example

Describe the different ways that seeds can be dispersed.

Some seeds, such as those of the dandelion, are dispersed by the wind. Other plants, such as the coconut, disperse their seeds in water. Yet others, such as burdock, produce seeds that stick to animals. And other plants, such as the geranium, eject their seeds.

Know >

1 What does dispersal of seeds mean?
2 Name a plant that disperses its seeds by the wind
3 What are burs?
4 What invention did George de Mestal develop after looking at burs?

Apply >>

5 Why are coconut palms found all over the world?

Extend >>>

6 What other inventions have been inspired by nature?

Enquiry >>>>

7 How many plants in your garden have seeds dispersed by wind, water, animals or ejection?
8 Collect some 'helicopter' seeds from your garden, park or school field. Plan an investigation to measure how far they travel during dispersal.

» Extend: Artificial reproduction in plants

For thousands of years, farmers and gardeners have been trying to produce plants with desirable characteristics. In the case of crops, this is usually a high yield. This means lots of large potatoes or apples. Yield is the useful part of any crop plant. For flowers this is often brighter colours that last longer or stronger smelling flowers.

Artificial sexual reproduction

Plant breeders are able to use the sexual reproduction of plants to artificially create new plants. These are artificially selected for the colour or smell of their flowers or their size and shape. The plant breeder will either collect pollen from one plant or remove the entire stamen containing pollen. This is then gently wiped onto the stigma of the second plant often using a fine paintbrush. The process of pollination then occurs naturally as described previously.

This is a costly and time-consuming process so is frequently used for expensive and desirable plants like orchids or roses. Importantly, this process produces new and unusual breeds of plant. If a high quality new breed is produced the breeder is given the chance to name the breed.

▲ This valuable orchid was produced by a plant breeder by the transfer of pollen from one plant to another

Artificial asexual reproduction

Plant breeders are able to use the asexual reproduction of plants to create many copies of the same plant. Unlike the artificial sexual reproduction described above, this process does not make new and unusual varieties of plants. It simply makes multiple copies of the parent plant. This plant was probably originally selected for its colour or smell of its flowers or its size and shape.

It is easy for plant breeders to collect the naturally produced plantlets of parent strawberry or spider plants and grow them into new plants for sale. However, not all plants produced plantlets in this way. Amazingly, some plants can grow from just a small number of cells from one parent. This process is called **tissue culture.** A small tissue sample is cut from the parent plant and placed into a culture medium containing water and minerals needed for healthy growth. The parent quickly repairs itself whilst the tissue culture grows into a new clone of the parent.

This is a cheap way of mass producing many copies of a desirable plant. Many of the plants you see in your local supermarket or garden centre have been grown in this way. Unlike, artificial sexual reproduction, plant breeders are far more likely to get similar plants to the parent.

▲ These orchids are clones of each other and have been produced by tissue culture

Tasks

1. What is yield?
2. What do some gardeners use to transfer pollen between plants?
3. Why do gardeners grow plants asexually?
4. What is tissue culture?
5. What characteristics are desirable in flowering plants?
6. How would a gardener help his plants reproduce sexually?
7. How would a gardener help his plants reproduce asexually?
8. Why are all cloned plants not identical?

Enquiry:
Which is the most effective helicopter fruit?

You learned earlier in this chapter that pollination occurs when a pollen grain lands on the stigma of a flower. A pollen tube is formed which transfers the pollen nucleus with its DNA to the ovary. A seed is then formed. You also learned that seeds can be dispersed (spread out) by wind, water, by latching onto animal fur or hair, or in their droppings. It is important that seeds to do not grow so close to their parent plant that they compete with it for resources like light and water.

Here we are going to look at the factors that affect the effectiveness of wind dispersion in detail. Many plants like the dandelion make very light seeds with structures that are easily caught in the wind. These can blow many miles to settle and grow into a new plant. Other seeds are heavier and don't usually travel so far. These plants have evolved a different way of moving their seeds away from the parent plants. They make seeds with helicopter-like wings that turn as they float through the air. Examples of this are the box elder, big-leaf maple, evergreen ash, tipu tree, empress tree and tree of heaven.

But what makes these seeds effective flyers? You will now investigate by making paper models of helicopters and testing their effectiveness. The longer a helicopter takes to drop to the floor the more successful it is.

The figure below shows a basic template for a paper helicopter. Your challenge is to change one factor of this template and investigate how this affects flight time. The factor you change will be the independent variable. The factor you measure will be the time it takes for the helicopter to fall to the floor in seconds using a stop watch. This is the dependant variable, because it depends upon the independent one (the variable you are changing).

Some of the independent variables you could change are the length or width of the wing, the length and width of the base, and whether adding paper clips to the base makes a difference. You need a sufficient spread in your independent variable so that you collect a range of results.

Remember it is important to only change one variable or your test will not be a fair test. You will need to standardise all other variables other than the one you change (the independent variable) and the one you measure (the dependent variable). These others are called control variables. So if you chose to investigate wing length, the wing width, base width and length and using only one paperclip must be standardised.

A crucial other fact that must be standardised is the height from which you drop your helicopter.

You will need to complete each drop several times to obtain three similar results. When this occurs your results are described as repeatable. At this stage you must ignore results that are not similar. After this you will need to calculate mean values of your three similar results. Prepare a table with space to record all measurements before you start.

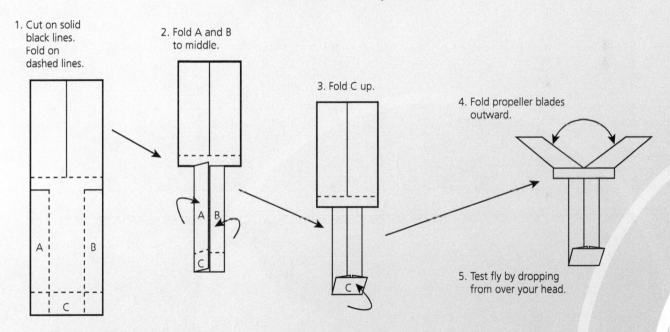

1. Cut on solid black lines. Fold on dashed lines.

2. Fold A and B to middle.

3. Fold C up.

4. Fold propeller blades outward.

5. Test fly by dropping from over your head.

▲ A model template for a paper helicopter

❶ In an experiment, what is the independent variable?
❷ In an experiment, what is the dependent variable?
❸ What is a non-similar result (i.e. a piece of data that does not fit the pattern) called?
❹ What do you do if your repeat result is not similar to previous ones?

Genes

Learning objectives

19 Variation

In this chapter you will learn...

Knowledge

- that there is variation between individuals of the same species
- that variation is inherited, environmental or a combination of both
- that variation between individuals is important for the survival of a species
- the definitions of the terms species, variation, continuous variation and discontinuous variation

Application

- how to explain whether characteristics are inherited, environmental or both
- how to draw bar charts or line graphs to show discontinuous or continuous variation
- how to explain how variation helps living organisms adapt to changing environments

Extension

- how to suggest the consequences of changes in the environment on species that live there
- how to explain why one species may adapt better than another to environmental change

20 Human reproduction

In this chapter you will learn...

Knowledge

- that the menstrual cycle prepares the female for pregnancy and stops if the egg is fertilised by a sperm
- that the developing foetus relies on the mother for oxygen and nutrients, to remove waste and to protect it against harmful substances
- that the menstrual cycle lasts approximately 28 days
- that if an egg is fertilised it often settles in the uterus lining

Application

- how to explain why some substances are passed from the mother to the foetus
- to use diagrams to describe the development of a foetus
- how to describe causes of low fertility in men and women
- how to describe key events on a diagram of the menstrual cycle

Extension

- how to explain why pregnancy is more likely at certain stages of the menstrual cycle
- how to explain how contraception and fertility treatments work
- how to predict the effect of cigarettes, alcohol or drugs on the developing foetus

19 Variation

» Transition: Selective breeding

Variation is another word for differences. The variation in a population of organisms is all the differences that these organisms have. These could have arisen naturally in wild animals or plants because of evolution. However, these differences could have been bred into them if they are pets, farm animals or crop plants.

Different breeds of dog can look very different. The Great Dane looks much larger than a Chihuahua. However, they are all one group (or species) of organisms. So any dog can breed with any other. You will learn more about this later in this chapter. So how do we have such different looking dogs? These differences did not come about naturally by evolution, but were selected by humans. We call this process **selective breeding**.

Around 30 000 years ago all dogs were originally wolves. At this time, some wolves started to live closer to humans. Slowly they became domesticated. This meant they started to live with and depend upon humans for the first time. Since this point we have selectively bred over four hundred different breeds. So how did this happen?

Your knowledge objectives:

In this chapter you will learn:
- that there is variation between individuals of the same species
- that variation is inherited, environmental or a combination of both
- that variation between individuals is important for the survival of a species
- the definitions of the terms species, variation, continuous variation and discontinuous variation

See page 233 for the full learning objectives.

★ You will learn about evolution in Pupil's Book 2, Chapter 19.

▲ Dogs and wolves have evolved from a common ancestor

If you were an early farmer many thousands of years ago you probably wanted a big dog to guard your sheep. Would you try to breed two smaller dogs together or two larger ones? Inheritance means you are more likely to get a larger dog from two larger parents. You will learn more about inheritance in Pupil's Book 2, Chapter 20. This was not nature selecting which organisms mate, but humans. So it is not natural selection which leads to evolution, but **artificial selection**.

Dogs have been selectively bred (by artificial selection) for a wide range of abilities or traits. Large dogs like German shepherds and rottweilers have been selectively bred to be guard dogs. Other breeds like the foxhound and bloodhound have been bred as hunting dogs. Some dogs like corgis were bred to herd cattle by nipping at their heels. Other dogs like golden retrievers have a gentle and caring nature and have been bred as pets or guide dogs. There is more variation in dog breeds as a result of selective breeding than many other species of animal that have evolved naturally.

▲ What have these breeds been artificially selected for?

Know >

1 What word describes variation?

2 What can all breeds of dog do?

3 When did wolves start to become domestic dogs?

4 Why were large dogs like German shepherds and rottweilers bred?

5 Why were dogs like the foxhound and bloodhound bred?

» Core: Causes of variation

Species

Key word

A **species** is a group of organisms that can interbreed to have fertile offspring.

A **species** is a group of organisms that can interbreed to have fertile offspring. All dogs are the same species, so all male and female dogs can sexually reproduce and produce fertile puppies. We call the differences between dogs 'breeds' because they are all one species.

A horse and a donkey are different species. They can have offspring that we call mules, but these are not fertile. So a mule cannot breed with a horse, donkey or another mule. Because of this horses and donkeys are different species. This also applies to lions and tigers. Ligers are infertile.

Variation

Key word

Variation is the differences within and between species.

Variation is both the sum of all the differences between:

- individuals within a species, or

- species in an ecosystem.

Variation between individuals within a species is called **intraspecific variation**. Examples of this include all the differences you see when you look at all the other people in your classroom or around your dinner table.

Variation between species in an ecosystem is called **interspecific variation**. Examples of this include the differences between lions and cheetahs on the plains of Africa. Both of these animals are top predators (at the top trophic level of food chains) and compete for food, water, territory and mates for reproduction.

The same is true of plants. There is intraspecific variation between the different oak trees in any woodland, but also interspecific competition between the oak trees and other species. All plants are producers (so at the bottom trophic level of all food chains) and compete for light, water, space and mineral salts.

There are two main causes of variation: genetic and environmental.

Genetic variation

Differences that have been caused by the genes we have inherited from our parents are called genetic variation. These differences were fixed the moment the sperm fertilised the egg and no changes can then be made.

Examples of genetic variation include:

- your sex (not gender which you can choose)
- eye colour
- blood group
- whether you have attached ear lobes or not
- whether you can roll your tongue or not.

▲ What examples of genetic variation can you see in this image?

Environmental variation

Differences that have not been caused by the genes we inherit but the environment we live in are described as environmental variation. These differences were not fixed when the sperm met the egg, but have developed later in life, not as a result of inheritance. Factors that could affect environmental variation include:

- the climate the organisms live in
- diet
- physical injury
- culture and lifestyle.

Examples of environmental variation include:

- dyed hair
- tattoos
- tanned skin
- scars.

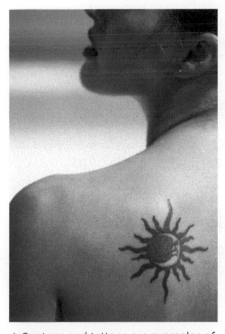

▲ Suntans and tattoos are examples of environmental variation

Genetic and environmental variation

Some of the more complicated differences we see in ourselves and other organisms are often caused by both genetic and environmental factors. Height and weight are good examples of these. Tall people tend to be born to tall families and so height is partly caused by genetic variation. However, if a person with the genes to be tall was not fed sufficiently as a child, their bones would not grow properly. Thus, height is partly caused by environmental variation.

Again, taller plants tend to have taller offspring. However, if any plant with genes for being tall did not have enough sunlight or water for photosynthesis, they would not grow sufficiently. So differences in plant height are also caused by both genetic and environmental variation.

Worked example

What is a species?

A group of organisms that can interbreed to have fertile offspring.

Know >

1 What is variation?

2 State an example of genetic variation.

3 State an example of environmental variation.

4 Why are horses and donkeys different species?

Apply >>

5 What type of variation exists between species of tree in a woodland?

Extend >>

6 What factors could affect environmental variation?

Enquiry >>

7 Complete a survey into examples of environmental and genetic variation in your class or your family.

Key words

A **continuous** variable has values that can be any number.

A **discontinuous** variable is one that is described in words or whole numbers.

>> Core: Types of data

When scientists look at variation they often complete a survey. The results from these give us two types of data called **continuous** and **discontinuous**.

Discontinuous variation

The results from surveys into discontinuous variation have values that come in groups or categories. Examples include:

- your sex (not gender which you can choose)
- eye colour
- blood group
- whether you can roll your tongue or not.

Do you recognise this list? Well done if you remembered from page 237 that these were the examples of factors caused by genetic variation. All these examples of genetic variation give discontinuous data in a survey. But the reverse is not true. Some examples of environmental variation also give us discontinuous variation. These include whether or not you have:

- dyed hair

- tanned skin

- tattoos

- scars.

The results of discontinuous variation are not able to be half way between groups. You cannot be half a male and half a female, or have half or one blood group and half of another. (Very occasionally some people have different eye colours.) You either have dyed hair or you don't.

We always draw results from discontinuous variation as a bar chart with gaps between the bars. We never draw a line of best fit on this type of bar chart because the data comes in groups and not a range.

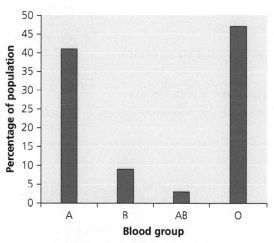

▲ The results from a survey into discontinuous variation shown as a bar chart

Continuous variation

The results from surveys into continuous variation have values that come in a range. Examples include:

- height

- mass

- hand span.

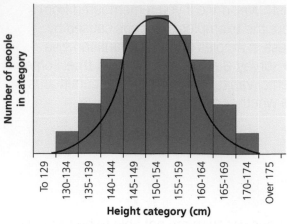

▲ The results from a survey into continuous variation shown as a line graph

In these examples you can be half way between any two values. So your height can be 170 cm or 171 cm, but it can also be 170.5 cm. Your mass can be 57 kg or 58 kg, but it can also be 57.5 kg.

We always draw results from continuous variation as a bar chart with the bars touching each other. A line of best fit is then added. This can be straight or curved. We do this because the data comes in a range and values can fall halfway between points.

Worked example

Why is mass an example of continuous variation?

Because the results obtained from a survey of mass are in a range rather than groups.

Know >

1 What is discontinuous variation?

2 What type of graph do you draw if your results are continuous?

3 What type of graph do you draw if your results are discontinuous?

Apply >>

4 An investigation into eye colour would lead to what type of data?

5 An investigation into height would lead to what type of data?

Extend >>

6 Will a survey of shoe size produce continuous data or discontinuous data?

Enquiry >>

7 Which other human features might give a graph like the one for height?

>> Core: Why is variation important?

Biodiversity is a measure of the differences between organisms within a species or the total number of species in an ecosystem. So in this regard it is the same as **variation**, which we said earlier in this chapter is both the sum of all the differences between individuals within a species, and also species in an ecosystem.

Here we will consider the difficulties that populations of organisms with lowered variation have. The difficulties that ecosystems with lowered biodiversity have were covered in Chapter 17.

Look around your classroom or family dinner table. What biodiversity can you see? You might see differences in height, weight, skin and hair colour, piercings, scars and tattoos. All these are examples of variation.

Now let's think about variation in other populations. How might a population of elephants differ? Some will be larger and stronger. Others will have better eyesight or remember where watering holes are more easily. Some will be more immune to illness or disease.

▲ Which of these two habitats has the greatest biodiversity?

Populations without much biodiversity are sometimes called inbred. Here the lack of differences between individuals means that they are less likely to be able to adapt to a change in their environment. This means they are more likely to become extinct.

Variation within a population is very important for its long-term survival. The small differences make any population able to withstand changes to its environment. With variation comes a wider gene pool. With this, populations are more easily able to adapt to changes like the introduction of a new predator or a disease.

Worked example

What is variation?

The sum of all the differences between individuals in a species, and also between all of the species in an ecosystem.

Know >

1 What is biodiversity?

Apply >>

2 How might the lions in a population be different from each other?

Extend >>

3 How do zoos reduce the effects of inbreeding?

» Extend: Evolution

Evolution is a process that explains how all life on Earth probably originated from an organism that we are all related to, called a common ancestor. The main points to know about evolution are:

★ **You will learn more about evolution in Pupil's Book 2, Chapter 19.**

- In every population there is variation.

- This must mean that some organisms are better adapted than others.

- These organisms are more likely to have offspring.

- Their offspring are more likely to have their parent's advantageous characteristics.

- Evolution occurs when these steps are repeated over many generations.

▲ Charles Darwin is called the father of evolution

Evolution is a never-ending process. It is always occurring. However, the rate of evolution may change. It is likely to be faster shortly after a large environmental change such as the asteroid impact that many think killed the dinosaurs, the ice ages and global warming. At these points, variation within populations is very important. Populations without sufficient variation are more likely to become extinct.

Horse's feet

Around 50 million years ago the horse was a small animal about the size of a dog. It had large feet to help it walk in wet marshland that made up much of its environment. Some individuals had larger feet.

This is a form of variation. They were more easily able to walk in the marshland. But other horses had smaller feet. They were lighter, and so more easily able to run away from predators but more likely to sink into the marshy ground.

Over the following thousands of years, the ground slowly began to become drier and grassier. The horses with the big feet no longer had the evolutionary advantage. They were now the ones that couldn't run fast enough on the dry ground to avoid predators. They were less likely to have offspring. The horses with smaller feet were now better adapted and were more likely to have offspring. In turn, these horses were more likely to have smaller feet. This process was repeated many, many times throughout the last 50 million years.

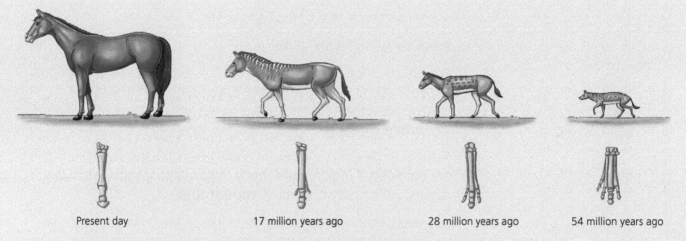

| Present day | 17 million years ago | 28 million years ago | 54 million years ago |

▲ Changes in the bone structure of horse's feet provide evidence for evolution

So now, the horse has evolved to be bigger, run faster and have relatively smaller feet because they became much better adapted to their changed environment. We can see these changes in the bone structure of fossilised horse's feet.

This also provides evidence for the importance of variation in populations and evolution. If some of the horses did not originally have smaller feet, the process of evolution may not have occurred and the horse could have become extinct. Variation is therefore a key part of the process of evolution, not just for the horse, but for all species of life on Earth.

Tasks

1. What is necessary for evolution to occur?
2. What term do we give to an organism that many species have evolved from?
3. What is more likely to occur to populations with less variation?
4. What environmental change occurred that resulted in the change in the size of horses' feet?
5. How is the horse's foot an example of evolution?

Enquiry:
Exploring continuous and discontinuous variation and survival

You learned earlier in this chapter that there are three causes of variation. These are:

- genetic, e.g. sex and eye colour

- environmental, e.g. tattoos and scars

- both, e.g. height and mass.

You also learned that the results from surveys fall into two groups depending upon the type of data that is recorded. These are continuous if the data comes in a range and discontinuous if it falls into discrete groups.

Many science experiments are completed until three similar results are obtained. If it didn't take many attempts to get three similar results, we can say our data is **repeatable**.

However, when we look at variation in ourselves or other species we often need to look at many more examples than in other science experiments. The more organisms that are investigated in a survey, the more confident we can be that our data is repeatable. Imagine if you completed a survey of your year group instead of your class. You would ask many more people and so could be many times more confident in your data.

Discontinuous variation

Choose an example to study that will give you discontinuous data. You could choose:

- your sex (not gender which you can choose)

- eye colour

- blood group

- whether you can roll your tongue or not.

Complete a survey using your family or friends. Before you start you might want to think about these points:

- What example will you choose?

- How many people will you ask?

- Will this be enough for you to be confident your data is repeatable?
- Will you draw a line graph or bar chart and why?
- Draw a table of your results before you start.

Completing a survey into continuous variation

Choose an example to study that will give you continuous data. You could choose:

- height
- weight
- hand span.

Complete a survey using your family or friends, thinking about the same questions as before.

Normal distribution in continuous variation

Many examples of biological surveys into continuous variation give us a very specific shape of line graph. You can see the results of this in the figure below.

We call this shape a **bell-shaped graph**. In it the majority of the organisms surveyed fall in the middle of the spread and there are fewer examples at either end. So if you looked at the height of all the people in your school year, most would be in the middle, and a much smaller number would be very short or tall.

We call the data that is presented in a bell-shaped graph **normally distributed**.

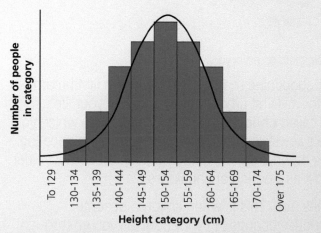

▲ The results from a survey into continuous variation showing the bell-shaped graph of normal distribution

1. What type of graph do you draw if your results are continuous?
2. What type of graph do you draw if your results are discontinuous?
3. What is continuous variation?
4. What is discontinuous variation?
5. If you sampled enough people in your height investigation what type of distribution would you see in your results?
6. Describe the shape of a graph with normal distribution.

20 Human reproduction

» Transition: The life cycle of a human

Your knowledge objectives:

In this chapter you will learn:
- that the menstrual cycle prepares the female for pregnancy and stops if the egg is fertilised by a sperm
- that the developing foetus relies on the mother for oxygen and nutrients, to remove waste and to protect it against harmful substances
- that the menstrual cycle lasts approximately 28 days
- that if an egg is fertilised it often settles in the uterus lining

See page 233 for the full learning objectives.

The genetic information that makes you was decided when your father's sperm met your mother's egg. As this point you were one fertilised egg cell, or zygote. Just one cell. Now you are probably over thirty million billion cells! That original cell copied itself many times in the first few weeks of your life as you settled into your mother's womb. From this point to around week 10 you were an embryo. You then developed into a foetus. By this time, you had developed to the point of being recognisably human. All your major body organs were present, if not fully developed.

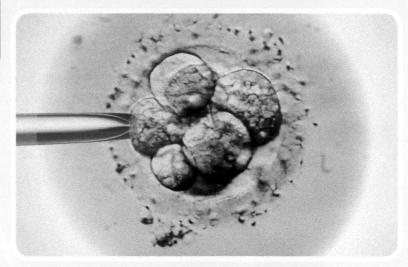

▲ A ball of cells developing into an embryo

You were born after about nine months. You took your first breath and probably cried for this first time. The next stage of your life began. This stage is called **childhood** and was a period in which your learned many of life's key skills such as moving and communicating. Childhood can be a challenging time for some people, but most older people look back on it as a period of happiness.

Adolescence comes next. This is when children become teenagers. You will notice changes in your body as you go through puberty. The average age for puberty in girls is 11. It is a year later in boys. It is important to remember that this is an average age. There are no set timetables and the usual range for puberty is between

8 and 14. Puberty lasts about four years and you will notice a number of changes during it. These are shown in the table below.

Girls	Boys
Pubic hair starts to grow on your body	Pubic hair starts to grow on your body
Breasts start to develop	Hair starts to grow on your face
Hips widen	Testicles start to produce sperm
Ovaries start to release eggs during the menstrual cycle (periods start)	A penis can become erect for this first time
Growth spurt	Growth spurt
	Voice breaks (becomes deeper)

After adolescence you will become an adult. At this time your body will be fully developed. You will be able to have children of your own. **Adulthood** is one of the longest stages of the human lifecycle. Regular exercise and healthy eating are important to stay fit and free from diseases.

The final stage of our life before death is called **old age**. This typically begins around retirement age when people stop working. This is often in the late 60s. Because our health care is much better, many more people are living much longer into their old age. This means that the average age of death is increasing. This poses challenges for governments all over the world.

Ethics

Saying when a new organism is a new life is quite difficult. Some people believe it occurs when fertilisation first occurs, so when the sperm meets the egg. Others think that a new life only begins when it can survive on its own. This means they believe a new organism is a new life many weeks after fertilisation. This is an ethical issue. This is one that some people disagree with for religious or moral reasons.

Know >

1 Where do babies develop?

2 When were you an embryo?

3 For how many months do babies develop before they are born?

4 What stage comes after childhood?

5 What is the average age of puberty for boys?

6 What is an ethical decision?

» Core: The male and female reproductive systems

All mammals reproduce in a similar way. We humans are mammals and so the way we reproduce is similar to other mammals like dogs, cows and elephants. All male mammals have **reproductive systems** that produces sperm. All female mammals have reproductive systems that produce eggs. When a male and female mammal mate, the sperm is introduced to the egg which it fertilises. This fertilised egg can then grow into a new life.

The male reproductive system

Sperm are the male sex cell. These contain half of the genetic information of the father. After puberty, men produce hundreds of thousands of sperm each day in their **testicles**. The testicles are glands so they also produce male sex hormones which affect the way in which a man's body develops.

When a man becomes sexually aroused, blood rushes to his **penis**. This makes it erect. It is now capable of passing in and out from a woman's vagina during sex. The end point of sex for men is an ejaculation. When this occurs hundreds of millions of sperm are forced from the testicles along sperm ducts. They are mixed with a fluid full of nutrients produced in glands. Together the fluid and sperm are called semen. This passes along the urethra in the penis and is ejaculated into the vagina.

The urethra can also carry urine from the body when a man urinates. The urine comes from the bladder where it is stored. Men are not able to urinate and ejaculate at the same time.

▲ Reproduction in humans is similar to other mammals like dogs, cows and elephants

> **Key word**
>
> A **reproductive system** includes all the male or female organs involved in reproduction.

> **Key words**
>
> A **testicle** is an organ in the male reproductive system where sperm are produced.
>
> The **penis** is the organ which carries sperm out of the male's body.

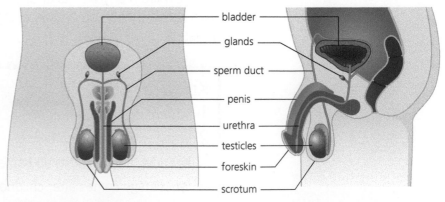

▲ The male reproductive system

The female reproductive system

Eggs are the female sex cell. These contain half the genetic information of the mother. These are produced in women's ovaries. Each **ovary** contains thousands of eggs. They are produced before women are born, unlike men who produce sperm from puberty onwards.

Ovaries are connected to the **uterus** or womb, by **oviducts**. These are sometimes called fallopian tubes. These are lined with tiny hairs called cilia, which move like a Mexican wave in a football match to propel the egg along the oviduct. **Fertilisation** of an egg by a sperm cell often happens in the oviducts.

The uterus is a muscular bag in which the fertilised egg settles and develops into a baby. The cervix is a ring or muscle at the bottom of the uterus that separates it from the **vagina**. The vagina is a muscular tube that leads from the outside of a women's body to the cervix. A man's penis passes in and out of the vagina during sex. The sperm are eventually ejaculated into the top of the vagina near the cervix. They swim through the cervix and uterus to reach the oviduct where fertilisation often occurs.

Urine does not pass through the vagina. Like men, it is stored in the bladder and passes along the urethra. However, it exits a woman's body just above the opening of the vagina.

Key words

An **ovary** is an organ which contains eggs.

The **uterus**, or womb is where a baby develops in a pregnant woman.

The **oviduct**, or fallopian tube, carries an egg from the ovary to the uterus and is where fertilisation occurs.

Fertilisation is the joining of a nucleus from a male and female sex cell.

The **vagina** is where the penis enters the female's body and sperm is received.

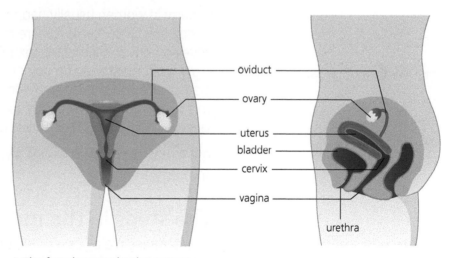

▲ The female reproductive system

Worked example

What are the roles of the male and female reproductive systems?

The male reproductive system produces sperm, which are ejaculated into the vagina of the female reproductive system. The female reproductive system produces eggs, allows fertilisation in the oviducts, and the development of a baby in the uterus.

Know >

1 What is the male sex cell?

2 What is the female sex cell?

3 What connects the ovaries to the uterus?

4 What two substances can the urethra carry in men?

Apply >>

5 What is the difference between sperm and semen?

Extend >>>

6 Make a model of a sperm cell and an egg cell.

Enquiry >>>>

7 Find out about fertilisation in animals, such as fish, which produce many more sperm and eggs. Why do they do this?

>> Core: The menstrual cycle

Puberty in women usually happens between ages 8 and 14. During this time, hormones are released which encourage the development of secondary sexual characteristics including the growth of breasts and the widening of hips. After puberty the menstrual cycle occurs approximately every twenty-eight days. This repeats until the menopause, after which women can no longer become pregnant. This is usually around age 50.

The menstrual cycle prepares a women's body to become pregnant and carry a baby.

The steps in the menstrual cycle

As with many biological cycles, the length of the stages in the menstrual cycle shows variation. For some women the following stages are regular, and for others they vary considerably.

Key facts

→ The menstrual cycle lasts approximately 28 days.
→ If an egg is fertilised it settles into the uterus lining.

Key words

Menstruation is the loss of the lining of the uterus during the menstrual cycle.

Ovulation is the release of an egg cell during the menstrual cycle, which may be met by a sperm.

Days	What happens
1 to 4	**Menstruation** occurs. Here the lining of the uterus from the previous menstrual cycle breaks down and passes out from the vagina. This is also called having a period. Usually between 5 and 12 teaspoons of blood pass from the vagina. During these days some women have stomach cramps and mood swings.
4 to 13	The lining of the uterus starts to rethicken in preparation for a baby.
14	A mature egg is released from an ovary. This is also called **ovulation**.
14 to 18	Fertilisation may occur if the woman has had sex. Sperm produced in male testicles may have been ejaculated from the penis into the vagina. Fertilisation will occur if one sperm can swim through the cervix and uterus to the oviduct to meet the egg.
14 to 28	If fertilisation occurs the lining of the uterus continues to thicken in preparation for the baby to develop.

If no fertilisation occurs, then the lining of the uterus remains thick until day 28. After this the cycle returns to day one and menstruation occurs again. |

Know >

1 When does puberty in girls occur?

2 What triggers puberty?

3 When in the menstrual cycle does menstruation happen?

4 After how many days does ovulation occur?

5 What are some secondary sexual characteristics in women?

6 What happens to the lining of the uterus if a woman becomes pregnant?

Apply >>

7 Why don't pregnant women have periods?

Extend >>>

8 Use the internet to find out the names of the two main hormones in the menstrual cycle.

Enquiry >>>>

9 Find out about the menstrual cycle of other animals. Is there a connection between their size and the length of their cycles?

» Core: Gestation and birth

Mammals carry their babies within them in a process called **gestation**. This is nine months in humans. The time varies in other mammals. For cats it is about two months and for elephants it is 21 months.

Mammals such as the kangaroo and koalas in Australia have a shorter gestation than others. After a shorter than normal time, the baby kangaroos and koalas climb a short distance to their mother's pouch. Here they see out the rest of their development.

During gestation the foetus is surrounded by **amniotic fluid** in the amniotic sac. This protects it from bumps and bruises as it develops.

During gestation the **foetus** receives the nutrients it needs from its mother. The baby is connected via its belly button through its **umbilical cord** to the mother's **placenta**. Here the baby's blood and that of its mother do not mix. They are separated by a partially permeable membrane which allows substances to diffuse between mother and child.

▲ A baby kangaroo is safe in its mother's pouch

> **Key words**
>
> **Gestation** is the process where the baby develops during pregnancy.
> **Amniotic fluid** is the liquid that surrounds and protects the foetus.
> A **foetus** is the developing baby during pregnancy.
> The **umbilical cord** connects the foetus to the placenta.
> The **placenta** is the organ that provides the foetus with oxygen and nutrients and removes waste substances.

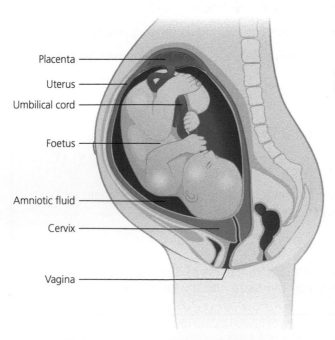

Placenta
Uterus
Umbilical cord
Foetus
Amniotic fluid
Cervix
Vagina

▲ The anatomy of an unborn baby and its mother

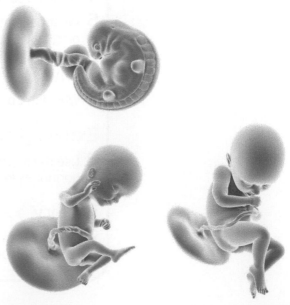

▲ The stages in development of a baby at 6, 16 and 36 weeks

★ **You will learn more about respiration in Pupil's Book 2, Chapter 17.**

Oxygen and glucose for respiration diffuse from the mother to the baby. Other nutrients needed for healthy growth also diffuse into the baby's blood. Sadly, other chemicals like nicotine from cigarettes and alcohol can diffuse into the baby. You will learn more about how these can affect the baby later in this chapter. Waste products like carbon dioxide diffuse backwards from the baby to the mother.

During the early weeks of gestation some women have morning sickness and cravings. These are often intense urges to eat unusual foods, often savoury or salty. Some people think this is the pregnant woman's body telling her it needs specific nutrients.

The process of birth is one of the most painfully difficult and amazing times any woman can have. Some births happen in a couple of hours, whilst others can take several days. The process usually begins by the pregnant mother's waters breaking. This occurs when the amniotic sac surrounding the baby bursts and releases the amniotic fluid which passes from the vagina.

▲ Pregnant women sometimes have unusual cravings

In the following hours the muscles of the uterus undergo a series of involuntary contractions. These move the baby from the uterus, through the narrow opening of the cervix to the vagina. The baby's head is the largest part and most babies are born head first to make this process easier.

Key word

The **umbilical cord** connects the foetus to the placenta.

For the seconds immediately after birth the baby is still getting its glucose and oxygen from its mother's placenta via the **umbilical cord**. Doctors quickly cut this and the baby takes it first ever breaths. The remainder of the placenta passes from the vagina shortly afterwards. This is often called the afterbirth.

Worked example

What is gestation?

The process in which a baby develops during pregnancy.

Know >

1 How long is gestation in humans?

2 What connects the mother to her baby?

3 What is often the first step in giving birth?

4 What are contractions?

Apply >>

5 What might happen to the baby if its umbilical cord became blocked?

Extend >>>

6 Put two eggs into two jam jars. Fill one with water, put the lids on and then shake both (see the photo below). What is this a model for?

Enquiry >>>

7 How long is gestation in other animals? Can you spot a link between gestation length and any other factor?

» Extend: Contraception and fertility

Contraception stops a woman becoming pregnant. There are many different forms of contraception but many types are called barrier contraceptives because they form a barrier between the sperm and eggs. Collectively sperm and eggs are called gametes or sex cells. Barrier contraception stops gametes joining.

Barrier contraception

Condoms are a form of barrier contraception. They are made from a rubbery latex which stretches around an erect male penis. Any ejaculated sperm are caught in a small nipple at the end of the condom and do not pass into the vagina. Condoms come in a variety of shapes, colours and sizes.

▲ Condoms

Other forms of barrier contraception include the female condom. This sits inside the vagina and works in the same way as the male condom.

Both the male and female condoms also have another important advantage. They stop the transfer of sexually transmitted infections (STIs). These include gonorrhoea, chlamydia and HIV/AIDS.

The diaphragm is another type of barrier contraception. These sit inside the vagina before the cervix and stop sperm passing through into the uterus. These do not stop the transmission of sexually transmitted infections.

▲ The contraceptive diaphragm

The contraceptive pill

Another commonly used form of contraception is the pill. There are several different types of pill but they contain the hormones oestrogen and progesterone. These hormones prevent eggs from being released from ovaries (ovulation). They also make it difficult for a sperm to reach an egg, and for the fertilised egg to implant in the uterus.

Women need to take the pill every day for 21 days and then stop for seven days. Some women find it easier to take a pill everyday so take a sugar pill during the seven days they don't need a proper pill. The pill needs to be taken at the same time each day. The pill can help reduce heavy or painful periods in some women.

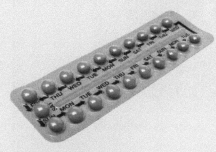

▲ The contraceptive pill

Some women find it difficult to take the pill at the same time each day. Others find the hormones within the pill have unwanted side effects such as mood swings and breast tenderness. To avoid taking the pill, some women have contraceptive implants. These are small devices that sit under the skin and release hormones into the blood as if the woman was taking the pill. They last for several years before needing to be replaced.

Fertility treatments

Becoming pregnant is not always easy. Women are most likely to become pregnant if they have sex around day 14 of the menstrual cycle. At this time an egg will have been released from an ovary and be travelling down the oviduct. Sperm released into the vagina at the start or the end of the menstrual cycle are very unlikely to result in a pregnancy. So many couples plan to have sex at specific times to maximise their chances. This can take some couples many months.

There are many different reasons why men and women are infertile (they cannot have children). The treatment that a person undergoes depends upon the reason why they are infertile.

▲ The process of *in-vitro* fertilisation

Some men do not produce enough mobile sperm to fertilise an egg in the oviduct. Some women do not regularly release eggs from their ovaries because their hormone levels are not suitably high enough. You learned about the hormonal control of the menstrual cycle earlier in this chapter. Women in this situation are given injections of hormones to help release eggs. If this does not work, or the issue is with the man's sperm, couples can have *in-vitro* fertilisation (IVF). *In vitro* means outside of the body. The woman is given an injection of hormones. Shortly afterwards several mature eggs are removed from one of her ovaries in a small operation. The eggs are mixed with sperm on a microscope slide and fertilisation occurs. On occasions the nucleus of the sperm containing the father's DNA is injected into the egg. At this point the doctor will select healthy looking fertilised eggs and they will be replaced into the uterus of the woman. If they embed properly into the lining of the uterus they will grow into a baby.

▲ Nadia Suleman had eight babies following a course of IVF

More than one fertilised egg is usually added to mean a greater chance of becoming pregnant. The most babies any known woman has born at one time is eight. Nadia Suleman underwent IVF and eight of the fertilised eggs settled into the lining of the uterus.

Tasks

1 Name two types of barrier contraception.
2 Which hormones are present in the contraceptive pill?
3 What happens during the 7 days in each menstrual cycle that women do not take the pill?
4 What advantage does a hormonal implant have over taking the pill?
5 What does 'IVF' stand for?

Enquiry:
The effects of drugs and alcohol on unborn babies

▲ The umbilical cord connects the unborn baby to its mother's placenta

★ **You will learn about respiration in Pupil's Book 2, Chapter 17.**

You learned earlier in the chapter that pregnant women carry babies for approximately nine months in a process called gestation. During this time, they are surrounded by amniotic fluid in the amniotic sac within the uterus to protect them. They are connected to their mother's placenta by the umbilical cord. The placenta is an amazing organ. It provides oxygen and nutrients from the mother to the baby.

The umbilical cord allows blood from the baby to flow up to the join between the cord and the placenta and then back to the baby. At this join, substances can be exchanged between the baby's blood and that of its mother. The blood of the baby and mother cannot actually mix.

For the whole of the gestation period the baby is dependent upon its mother for the substances it needs to grow. The baby's cells require glucose and oxygen to complete respiration to release energy for growth. The baby also needs smaller amounts of other substances like vitamins and minerals for healthy growth.

These substances are not forced from the mother's blood to the baby's blood. They move from an area of high concentration in the placenta to an area of lower concentration in the baby's blood or the other way around. This process is called diffusion. It is not controlled. It happens until the concentrations of the two areas are equal.

Substance	Where originally in high concentration	Where originally in low concentration	So it moves from... to...
Oxygen	In the mother's blood from her lungs	In the baby's blood because its cells have been respiring	Mother to baby
Glucose	In the mother's blood from her diet	In the baby's blood because its cells have been respiring	Mother to baby
Vitamins and minerals	In the mother's blood from her diet	In the baby's blood because they have been used for healthy growth	Mother to baby
Carbon dioxide	In the baby because its cells have been respiring	In the mother's blood because she has been breathing it out from her lungs	Baby to mother

Unfortunately, it is not just useful substances that can cross from the mother to the baby through the placenta. Other harmful substances can move across as well. These include alcohol and many of the chemicals in cigarettes. Because of this, many pregnant women choose to give up drinking and smoking. Taking illegal drugs can harm unborn babies. Stimulants such as cocaine can affect the blood flow to the baby. Cannabis can cross the placenta like tobacco, directly affecting the baby. Babies whose mother's took drugs while pregnant are often smaller, have an increased risk of miscarriage or being stillborn (being born dead), and a slower development after being born.

Chemical	Effect on unborn baby
Alcohol	Higher chance of miscarriage or premature birth, low birth weight and other issues like learning difficulties and behavioural problems
Nicotine from cigarettes	Higher chance of miscarriage or premature birth, low birth weight, higher chance of stillbirth and cot death
Caffeine	Higher chance of miscarriage and low birth weight

▲ What damage is this mother doing to her unborn baby?

The NHS also recommends that pregnant women think about reducing or stopping consuming blue cheeses, raw eggs, unpasteurised milk, pate, raw meat, fish, alcohol and caffeine because of the effects they can have on unborn babies.

1 What is gestation?
2 What connects the mother to her baby?
3 What bad substances can cross the placenta and harm the baby?
4 What surrounds the baby and protects it?
5 What does the baby receive from the mother in the placenta?
6 What does the mother receive from the baby in the placenta?
7 What foods does the NHS recommend pregnant women avoid?

Glossary

A

When light is **absorbed** by a material, energy is transferred to it.

During **absorption,** energy is transferred to a material.

Acceleration How quickly an object's speed increases or decreases.

An **acid** is a solution which has a pH of less than 7.

Air pressure is caused by the weight of the atmosphere above an object or surface.

An **alkali** is a solution of a base. It has a pH which is greater than 7.

An **alloy** is a mixture of a metal with another element, which is usually a metal.

Amniotic fluid is the liquid that surrounds and protects the foetus.

The height of the wave, measured from the middle, is called the **amplitude**.

The **angle of incidence** is between the normal and the incident ray.

The **angle of reflection** is between the normal and the reflected ray.

An **antagonistic muscle pair** is when muscles are working in unison to create movement.

The average **auditory range** for an animal is the difference between the highest and lowest frequency it can hear.

Average speed The overall distance travelled divided by overall time for a journey.

B

A **base** is a chemical which reacts with an acid to neutralise it.

Bioaccumulation is the build-up of toxic chemicals in organisms found at higher trophic levels.

Biodiversity is the variety of living things. It is measured as the differences between individuals of the same species, or the number of different species in an ecosystem.

Biomass is the dry mass of any organism.

Boiling is when a liquid turns into a gas when it is heated above its boiling point.

Bone marrow is the tissue found inside some bones where new blood cells are made.

A substance is **brittle** if it shatters when it is hit with a hammer, or if it breaks when you try to bend it.

C

Cancer is a group of diseases that involve cells growing and multiplying out of control. The cancer may spread to other parts of the body.

The **carpel** is the female part of the flower, made up of the stigma (where the pollen lands), style and ovary.

Cartilage is the smooth tissue found at the end of bones, which reduces friction between them.

A **celestial body** (or astronomical body) is an object in space. Planets, moons, stars, asteroids and comets are all examples.

A **cell** is the simplest unit of a living organism and it contains parts which carry out the seven life processes.

Charges are tiny particles in wires and components.

A **chemical store** of energy is emptied during chemical reactions.

Chromatography can be used to separate different coloured substances in a mixture.

A **circuit** connects components such as cells and lamps so charges can move.

A **compound** is a pure substance made from two or more elements which are chemically combined in a fixed ratio of atoms.

A **concave** lens is thinner in the middle and spreads out light rays.

Condensing is when a gas turns into a liquid when it is cooled.

A **consumer** is an animal that eats other animals or plants.

A **continuous** variable has values that can be any number.

A **convex** lens is thicker in the middle and bends parallel light rays towards each other.

A **correlation** is a relationship between two variables. The dependent variable increases or decreases as the independent variable increases.

The flow or movement of charge is called **current** and is measured in amperes (A).

D

A **decomposer** is an organism that breaks down dead plant and animal material so nutrients can be recycled back to the soil or water.

Density is the mass of a substance divided by its volume. Density may have units of kg/m^3, or g/cm^3.

The **dependent** variable is the factor you measure in an investigation.

Diffusion is when one substance spreads out from a high concentration, through a gas or a liquid, due to the random movement of particles.

A **discontinuous** variable is one that is described in words or whole numbers.

The temporary movement of a medium as the wave travels is called **displacement**.

In a **displacement reaction**, a more reactive element will replace a less reactive element in a compound.

When energy stores are filled or emptied, some energy is **dissipated** and becomes shared between more energy stores. This reduces the useful energy available.

Dissolving is when a substance mixes completely with a liquid called a solvent.

Distillation is a method of separating a mixture by boiling and then condensing the gas.

A substance is **ductile** if it can be drawn into wires.

E

A sound reflecting from a surface to a detector or listener is called an **echo**.

The **ecliptic** is the plane in which the planets orbit the Sun.

An **ecosystem** is the living things in a given area and their non-living environment.

An **elastic store** of energy is filled when a material is stretch or compressed.

An **electrical conductor** is very easy for charges to move through.

An **electrical insulator** is very hard for charges to move through.

Electrolysis is when a liquid is broken down into elements using electricity.

In an **electrostatic field** any charged object or particle experiences a force.

An **element** is a pure substance made from only one type of atom. Elements are listed on the periodic table.

An **energy resource** is something with stored energy that can be released in a useful way.

The **environment** is the surrounding air, water and soil where an organism lives.

Enzymes are proteins made by living organisms to speed up chemical reactions that take place within cells.

The **equator** is an imaginary line that runs around the Earth and separates the northern and southern hemispheres.

Erosion is the movement of rock by water, ice or wind.

When you **evaluate** something, you consider the advantages and disadvantages and come to a decision which you can justify.

Evaporation is when a liquid slowly turns into a gas at a temperature that is below its boiling point. Molecules escape from the surface when they have enough energy.

An **exoplanet** is a planet that orbits a star outside our solar system.

F

Fertilisation is the joining of a nucleus from a male and female sex cell.

The area in which an object feels a force is called a **field**.

Filtrate is the liquid that passes through a filter.

Filtration is the way to separate an insoluble solid from a liquid.

A **foetus** is the developing baby during pregnancy.

A **food chain** is part of a food web, starting with a producer, ending with a top predator.

A **food web** shows how food chains in an ecosystem are linked.

Fossils are the preserved remains or traces of living things from thousands or millions of years ago.

Fossil fuels are non-renewable energy resources formed from the remains of ancient plants or animals. Examples are coal, crude oil and natural gas.

A **fraction** is a liquid obtained from fractional distillation. If it is a pure liquid, it will have a specific boiling point. If it is a mixture of liquids, their boiling points will be similar.

Fractional distillation is the method used to separate a mixture of liquids which have different boiling points.

Freezing is when a liquid turns into a solid as it cools down.

The number of complete waves detected in one second is called the **frequency**.

A **fruit** is the structure that the ovary becomes after fertilisation, which contains seeds.

Fuels are substances which can be burnt to release energy.

G

A **galaxy** is a group of stars which are held together by gravity and their orbiting planets. Each galaxy contains billions of stars.

A **gamete** is a sex cell. The male gamete (sex cell) in animals is a sperm, and the female gamete is an egg.

Gas pressure is the force pushing outwards on the walls of a container, which is caused by the gas particles colliding with the container walls.

In the **geocentric model** of the solar system, all planets and stars orbit the Earth.

Gestation is the process where the baby develops during pregnancy.

The **gradient** of a graph is the slope. It is usually found by dividing the vertical change by the horizontal change.

A **gravitational store** is filled when an object is raised. This is also known as gravitational potential energy.

H

In the **heliocentric model** of the solar system, all planets orbit the sun.

A **heliograph** is a device that uses sunlight to create a signal.

A **hypothesis** is a prediction which can be tested by experiments or observations.

I

Igneous rocks are formed from molten rock.

An **impurity** is a substance present in small quantities in a substance that would otherwise be pure.

Incandescent means something is glowing as a result of being heated up. In light bulbs a metal filament is electrically heated until it emits light.

The incoming ray is called the **incident ray**.

The **independent variable** is the factor you change in an investigation.

An **indicator** is a chemical which can be used to identify whether a substance is acidic or alkaline because it changes colour.

A substance is **insoluble** in a particular liquid if it will not dissolve into the liquid.

Invasive species are species of animal or plant that have been introduced to new ecosystems by humans. They have negative effects on the other native species.

J

Joints are the points at which bones meet.

K

A **kinetic store** of energy is filled when an object speeds up. This is also known as kinetic energy.

L

Ligaments connect bones in joints.

A **light year** is the distance that light travels in one Earth year.

Limitations are the weaknesses of a model.

The displacement of a **longitudinal wave** is along or in line with the direction of wave travel.

A **luminous** object is a source of light rather than reflecting or scattering light from somewhere else.

M

A substance is **malleable** if it can be beaten into a new shape.

Mass is the amount of stuff (matter) in an object. Mass is measured in kilograms (kg).

The **mean** of a set of data is calculated by adding together the values and then dividing by the number of values which were added.

Melting is when a solid turns into a liquid when it is heated.

The **melting point** of a solid is the temperature at which it melts when it is heated.

Menstruation is the loss of the lining of the uterus during the menstrual cycle.

Metals are a group of substances that are all good conductors of electricity and heat. They are usually shiny, malleable and ductile, and almost all are solid at room temperature.

Metamorphic rocks are formed from other rocks, due to the action of heat and pressure.

Minerals are solid chemical compounds that are found in the Earth's crust. Because each mineral is a compound, it will have a specific chemical formula.

A **mixture** is when two or more elements or compounds are mixed together but have not chemically reacted.

A **model** in science doesn't have to be a small scale representation of a larger object, like a model aeroplane is a small version of an aeroplane. We use the term 'model' when we have developed a way to explain something tricky.

Molecules are tiny particles of a compound. Molecules are made from atoms which are strongly bonded together.

A **multi-cellular** organism is made from many cells. All animals and plants are multi-cellular organisms.

N

Electrons are tiny, **negatively charged** particles that are part of atoms.

A **neutral** solution has a pH of 7 and is neither acidic nor alkaline.

Sound that is unwanted or annoying is sometimes called **noise pollution**.

Non-contact force is a force that acts without direct contact.

Non-metals are elements, compounds or mixtures that are not metals. They can be solids, liquids or gases, and have a range of properties.

A **non-renewable** energy resource is one that cannot be replaced and will eventually be used up. Examples are coal, crude oil and natural gas.

The **normal** is an imaginary line at right angles to the surface. Scientists measure light angles *from* the normal.

O

An **opaque** material allows *no* light to pass through it.

To **orbit** an object in space means to circle around it, due to the attraction of gravity.

The **orbit** of a planet, moon or satellite is the path that it takes around a larger body.

An **organ** is a group of different tissues which work together to carry out a job.

An **oscilloscope** displays electrical signals on a screen. Often these electrical signals were converted from sound waves being detected by a microphone.

An **outlier** is a piece of data that does not fit the pattern.

An **ovary** is an organ which contains eggs.

The **oviduct**, or fallopian tube, carries an egg from the ovary to the uterus and is where fertilisation occurs.

Ovulation is the release of an egg cell during the menstrual cycle, which may be met by a sperm.

Ovules are the female sex cells in plants found in the ovary.

An **oxidation reaction** is when a substance reacts with oxygen.

P

Components on different loops are said to be **parallel** to each other.

A **particle** is a very tiny object which is too small to be seen, even with a microscope.

The **penis** is the organ which carries sperm out of the male's body.

The time taken for one complete wave to pass a point is called the **period**.

Photosynthesis is the process that uses the Sun's energy to convert carbon dioxide and water into glucose and oxygen.

The **pH scale** is the way we measure whether a solution is acidic or alkaline.

A whistle or squeak is a sound with a high **pitch**. Thunder has a low pitch.

The **placenta** is the organ that provides the foetus with oxygen and nutrients and removes waste substances.

Pollen contains the plant male sex cells found on the stamens.

Pollination is the transfer of pollen from the male part of the flower to the female part of the flower on the same or another plant.

A **population** is a group of the same species living in an area.

An object which has lost electrons will be **positively charged**. An object which has gained electrons will be negatively charged.

Potential difference is the amount of energy shifted from the cell to each unit charge, or from each unit charge to a component, in volts (V). Sometimes called voltage.

Scientists make **predictions** about what will happen in an experiment based on experience and scientific ideas.

Power is how fast energy is transferred by a device. It is measured in watts.

A **producer** is a green plant or algae that makes its own food using sunlight.

A **property** of a substance is a word that can be used to describe how it behaves. Examples of properties are: high melting point, good conductor of electricity, hard, strong, etc.

A **pure** substance contains only one type of element or compound.

R

The **range** of a data set is the difference between the smallest and largest measurement.

Reactivity is how easily a substance reacts with other chemicals.

A **reactivity series** shows which substances in a group are most and least reactive.

There is a **real difference** between two means if the range of the data sets does not overlap much.

The outgoing ray is the **reflected ray**.

Refraction is a change in the direction of light that happens when it goes from one material into another. The new direction, measured from the normal, is the angle of refraction.

Relative motion Different observers judge speeds differently if they are in motion too. An object's speed is relative to the observer's speed.

An energy resource that can be replaced as it is used, and will not run out is **renewable**. Examples are solar, wind, waves, geothermal and biomass.

A set of results is **repeatable** if the repeat readings are close together. This means the range is small.

A **reproductive system** includes all the male or female organs involved in reproduction.

The **residue** is the solid that is collected in the filter paper.

Resistance measures how hard it is to push charges through a material or component. It is measured in ohms (Ω).

The **retina** is a layer of light detecting cells at the back of the eye where an image is formed.

The **rock cycle** describes how different rocks are made and changed over long periods of time.

Rocks are mixtures of minerals found in the Earth's crust.

S

A **salt** is a chemical produced when an acid is neutralised. Part of the salt comes from the acid, and the other part is almost always a metal.

Scattering is when light bounces off an object or surface in all directions, not just one.

A **secondary source** is a source of data which was not collected by your own experiment. This could include a data book, or someone else's research.

Sedimentary rocks are formed when small particles settle out from slow moving water and then get squashed and glued together.

A **seed** is the structure that contains the embryo of a new plant.

Components are in **series** if they are in the same loop.

Solubility is the maximum mass of a solute that will dissolve in a specific volume of solvent.

A substance is **soluble** in a particular liquid if it dissolves in it to make a solution.

A **solute** is a substance (normally a solid) which can dissolve into a solvent.

A **solution** is a mixture made from a liquid and a substance dissolved into it (usually a solid).

A **solvent** is a substance (normally a liquid) which can dissolve another substance.

A **species** is a group of organisms that can interbreed to have fertile offspring.

Speed How much distance is covered in a certain time.

A **star** is a celestial body which gives out light. A star may have a solar system of planets.

Strata are layers within a sedimentary rock.

Structural adaptations are features of a cell which allow it to do its job.

Some substances do not commonly exist as a liquid and **sublime** straight from a solid to a gas when heated.

A **system** is an object, or a group of objects, that interact. We can choose what we include in the system.

T

Tendons connect muscles to bones.

A **testicle** is an organ in the male reproductive system where sperm are produced.

A **thermal store** of energy is a measure of the energy stored in a substance due to the vibration and motion of particles. It is sometimes just called thermal energy.

A **tissue** is a group of cells of one type.

A **translucent** material allows *some* light to pass through it.

If light is perfectly **transmitted** by a material it goes through it without any energy being absorbed.

A **transparent** material allows *all* light to pass through it.

A **trophic level** is a stage in a food chain or web.

U

Ultrasound is the name given to sound waves with frequencies higher than the human auditory range.

The **umbilical cord** connects the foetus to the placenta.

A **uni-cellular** organism is a living thing made form just one cell.

The **universe** is all of space (and time).

The **uterus**, or womb is where a baby develops in a pregnant woman.

V

A **vacuum** is a space with no particles in it, so sound waves cannot travel through it.

The **vagina** is where the penis enters the female's body and sperm is received.

A **variable** is something that can be changed, measured or controlled.

Variation is the differences within and between species.

A **vibration** is a back and forth motion that repeats in a pattern.

The **volume** of a sound, measured in decibels (dB) measures how loud it is.

W

Weathering is the wearing down of rock by physical, chemical or biological processes.

Weight is the force of gravity acting on an object. Weight is measured in Newtons (N).

INDEX

Photo credits

The Publisher would like to than the following for permission to reproduce copyright material:

p.4 © Matt Gibson/veneratio/Fotolia; **p.8** *t* © Olivier Morin/AFP/Getty Images, *b* © DBurke/Alamy; **p.9** © Keith Larby/Alamy Stock Photo; **p.12** © Matt Gibson/veneratio/Fotolia; **p.14** © NASA; **p.16** © Ed Brown/Alamy Stock Photo; **p.17** *tl* © Google, Inc., *tr* © The BLOODHOUND Project, *b* © Google, Inc.; **p.20** © Ruth Jenkinson/Getty Images; **p.21** *t* © Georgios Kollidas/Fotolia, *b* © NASA; **p.22** © Anne Abouchar 2013; **p.23** *t* © Xinhua/Alamy Stock Photo, *b* © chartcameraman/Fotolia.com; **p.24** © C Davenport; **p.25** © Sipa Press/Shutterstock/REX; **p.26** © NASA; **p.28** © Pavel Sytilin/Fotolia; **p.30** © Doug Martin/Science Photo Library; **p.31** *l* & *r* © Image Source/Getty Images; **p.32** © Pavel Sytilin/Fotolia; **p.33** *l* © Perutskyi Petro/Shutterstock, *r* © GIPhotoStock/Science Photo Library; **p.34** © Adam Hart-Davis/Science Photo Library; **p.38** © Ian Horsewell; **p.43** © Patrick Dumas/Look At Sciences/Science Photo Library; **p.44** © Ted Kinsman/Science Photo Library; **p.45** © Peter Menzel/Science Photo Library; **p.46** © EtiAmmos/Shutterstock.com; **p.47** © Science Source/Science Photo Library; **p.48** © Broker/Fotolia; **p.50** © Rosemary Roberts/Alamy; **p.51** © C Davenport; **p.52** *tl* © Adriano Pecchio/iStock/Thinkstock, *ctl* © U Gernhoefer/Fotolia, *ctr* © Layne Kennedy/Getty Images, *cbl* © lightpoet/Fotolia, *cbr* © Kelpfish/Fotolia.com; **p.54** *t* © Berca/Fotolia.com, *b* © broker/Fotolia.com; **p.56** *tl* © Sommersby/Fotolia.com, *tr* © Elenathewise/Fotolia.com, *b* © Sheila Terry/Science Photo Library; **p.59** © Željko Radojko/Fotolia; **p.60** *t* © Fotolia, *b* © Photoshot; **p.62** © Mino Surkala/Shutterstock.com; **p.63** © Joe Gough/Fotolia.com; **p.64** © Kwame Zikomo/Purestock/Alamy Stock Photo; **p.65** © Anton Kudelin/Shutterstock.com; **p.66** © Johanna Goodyear/Fotolia; **p.67** *t* © Tony McConnell/Science Photo Library, *b* © GIPhotoStock/Science Photo Library; **p.69** © lzf/Shutterstock.com; **p.70** *t* © Middle Temple Library/Science Photo Library, *b* © Adam Hart-Davis/Science Photo Library; **p.71** *t* © Dr Gary Settles/Science Photo Library, *b* © Fenton/Fotolia; **p.72** © Jose Manuel Gelpi/Fotolia; **p.74** © Elnur/Shutterstock.com; **p.75** *tl* © Science Photo Library, *tr* © sciencephotos/Alamy Stock Photo, *bl* © Corbis/Education, *br* © Image Source/Getty Images/Secondary School IE371; **p.76** *tl* © Allan Cash Picture Library/Alamy Stock Photo, *tr* © Andrew Lambert Photography/Science Photo Library, *b* © Martyn F Chillmaid/Science Photo Library; **p.80** *tl* © dutchpilot22/Fotolia, *tr* © Darren Baker/Fotolia, *bl* © broker/Fotolia.com, *br* © Imagestate Media (John Foxx)/Vol 09 Lifestyles Today; **p.81** *tl* © auremar/Fotolia.com, *bl* © stocksolutions/Shutterstock.com, *br* © Eric Cabanis/AFP/Getty Images; **p.82** *t* © Eleonora_os/Shutterstock.com, *b* © GIPhotoStock/Science Photo Library; **p.84** © George Silk/The LIFE Picture Collection/Getty Images; **p.86** © Jose Manuel Gelpi/Fotolia.com; **p.88** *l* & *r* © 1000 Words/shutterstock.com; **p.89** *tl* & *tr* © ER Degginger/Science Photo Library, *b* © GIPhotoStock/Science Photo Library; **p.90** © Alberto Pomares/iStockphoto.com; **p.92** © Trevor Clifford Photography/Science Photo Library; **p.94** *l* © kavring/Fotolia, *m* © Photodisc/Getty Images/World Landmarks & Travel V60, *r* © kichatof/Fotolia.com; **p.95** © Neamov/Shutterstock.com; **p.100** *t* © Cebas/Shutterstock.com, *m* © Neil Dixon, *b* © GustoImages/Science Photo Library; **p.102** *tl* © John Mainstone, University of Queensland via Wikipedia Creative Commons (https://creativecommons.org/licenses/by-sa/3.0/deed.en), *ml* © Neil Dixon, *mr* © Neil Dixon, *b* © Neil Dixon; **p.103** *t* © Natural History Museum, London/Science Photo Library, *b* via Wikipedia Creative Commons (https://creativecommons.org/licenses/by-sa/3.0/deed.en); **p.106** *l* © jubax/Fotolia.com, *r* © Fotolia; **p.107** *l* & *r* © Trevor Clifford Photography/Science Photo Library; **p.108** *l* © zayatssv/Fotolia, *ml* © Imagestate Media (John Foxx)/Vol 13 Backgrounds & Structures, *mr* © Imagestate Media (John Foxx)/SpIrit V3070, *r* © Africa Studio/Shutterstock.com; **p.112** © Science Photo Library; p.114 © Trevor Clifford Photography/Science Photo Library; **p.117** © Leah-Anne Thompson/Shutterstock.com; **p.118** © GIPhotoStock/Science Photo Library; **p.120** *tl* © Han van Vonno/Fotolia, *tr* © evgenyatamanenko/Fotolia, *bl* © Dmitry Vereshchagin/Fotolia, *br* © Fatbob/Fotolia; **p.121** *l* © Westend61 GmbH/Alamy Stock Photo, *r* © Cultura RM Exclusive/Peter Muller/Getty Images; **p.122** *t* © Happy Helfer/Fotolia, *b* © Mark Weiss/Getty Images; **p.124** © Martyn F Chillmaid/Science Photo Library; **p.126** © Andrew Lambert Photography/Science Photo Library; **p128** © GIPhotoStock/Science Photo Library; **p.131** © VDWI Automotive/Alamy Stock Photo; **p.133** *t* © Martyn F Chillmaid/Science Photo Library, *b* © Andrew Lambert Photography/Science Photo Library; **p.134** *t* © Getty Images/iStockphoto/Thinkstock, *b* © Phil Degginger/Alamy Stock Photo; **p.136** *l* & *r* © Neil Dixon; **p.137** © Neil Dixon; **p.138** © gsplanet/Shutterstock.com; **p.140** © Martyn F Chillmaid/Science Photo Library; **p.142** © Trevor Clifford Photography/Science Photo Library; **p.145** © antoniodiaz/Shutterstock.com; **p.146** © Joe Gough/Fotolia; **p.148** *tl* © JMB Photography/Fotolia, *tr* © April Turner/iStockphoto.com, *b* © BarryTuck/Shutterstock.com; **p.149** *t* © Jorge Salcedo/Shutterstock.com, *b* © Dirk Wiersma/Science Photo Library; **p.150** © Sputnik/Science Photo Library; **p.152** *t* © I love photo/Shutterstock.com, *b* © Getty Images/iStockphoto/Thinkstock; **p.153** © TMAX/Fotolia.com; **p.154** *l* © Dr. Marli Miller/Visuals Unlimited, Inc./Science Photo Library, *m* © silky/Shutterstock.com, *r* © Joe Gough/Fotolia; **p.155** © encikarel/Shutterstock.com; **p.159** © Michael Szoenyi/Science Photo Library; **p.160** © Detlev Van Ravenswaay/Science Photo Library; **p.161** © Neil Dixon; **p.162** © magann/Fotolia; **p.164** © Mark Garlick/Science Photo Library; **p.166** *t* © Imagestate Media (John Foxx)/Computers & Hi-tech SS08, *b* © Eckhard Slawik/Science Photo Library; **p.167** © NASA/Science Photo Library; p.169 *t* © Mark Garlick/Science Photo Library, *b* © NASA/Science Photo Library; **p.170** *t* © Neil Dixon, *b* © Thinkstock/Getty Images/Hemera; **p.171** © British Antarctic Survey/Science Photo Library; **p.172** © American Philosophical Society/Science Photo Library; **p.173** *l* © Paul D Stewart/Science Photo Library, *r* © Sheila Terry/Science Photo Library; **p.174** © iwikoz6/Fotolia; **p.176** *t* © Eon Neal/Staff/Getty Images, *m* © Will & Deni McIntyre/Science Photo Library, *bl* © Sailorr/Fotolia.com, *br* © tbel/Fotolia.com; **p.177** © JohanSwanepoel/Fotolia; **p.178** © Liya Graphics/Shutterstock.com; **p.179** *t* © James Cavallini/Science Photo Library, *b* © Rob van Esch/Shutterstock.com; **p.186** *l* © ScreenProd/Photononstop/Alamy Stock Photo, *m* © Entertainment Pictures/Alamy Stock Photo, *r* © BrendanHunter/Getty Images; **p.187** *tl* © Romanchuck/Fotolia, *tm* © studio306fotolia/Fotolia.com, *tr* © Oleksandr Dibrova/Fotolia, *m* © Museum of Innovation & Science, *b* © John B Carnett/Contributor/Getty Images; **p.188** © Aurora Photos/Alamy Stock Photo; **p.190** *t* © Equinox Graphics/Science Photo Library, *b* © Kirill Kurashov/Fotolia; **p.194** *t* © Juergen Berger/Science Photo Library, *m* © David M. Phillips/Science Photo Library, *b* © RomanenkoAlexey/Shutterstock.com; **p.195** *t* © Thomas Deerinck, NCMIR/Science Photo Library, *b* © Eye Of Science/Science Photo Library; **p.196** *t* © Eye Of Science/Science Photo Library, *m* © Dr Keith Wheeler/Science Photo Library, *b* © Steve Gschmeissner/Science Photo Library; **p.197** © Dr David Furness, Keele University/Science Photo Library; **p.200** *t* © St Mary's Hospital Medical School/Science Photo Library,